14. Band, 1. Heft

Fortschritte der chemischen Forschung
Topics in Current Chemistry

H. A. Bent   Localized Molecular Orbitals and Bonding in
Inorganic Compounds........................ 1

W. D. Ehmann   Non-Destructive Techniques in Activation Analysis   49

R. B. King   The Fragmentation of Transition Metal Organometallic Compounds in the Mass Spectrometer... 92

Herausgeber:

Prof. Dr. A. Davison   Department of Chemistry, Massachusetts Institute
of Technology, Cambridge, MA 02139, USA

Prof. Dr. M. J. S. Dewar   Department of Chemistry, The University of Texas
Austin, TX 78712, USA

Prof. Dr. K. Hafner   Institut für Organische Chemie der TH
6100 Darmstadt, Schloßgartenstraße 2

Prof. Dr. E. Heilbronner   Physikalisch-Chemisches Institut der Universität
CH-4000 Basel, Klingelbergstraße 80

Prof. Dr. U. Hofmann   Institut für Anorganische Chemie der Universität
6900 Heidelberg 1, Tiergartenstraße

Prof. Dr. K. Niedenzu   University of Kentucky, College of Arts and Sciences
Department of Chemistry, Lexington, KY 40506, USA

Prof. Dr. Kl. Schäfer   Institut für Physikalische Chemie der Universität
6900 Heidelberg 1, Tiergartenstraße

Prof. Dr. G. Wittig   Institut für Organische Chemie der Universität
6900 Heidelberg 1, Tiergartenstraße

Schriftleitung:

Dipl.-Chem. F. Boschke   Springer-Verlag, 6900 Heidelberg 1, Postfach 1780

Springer-Verlag   6900 Heidelberg 1 · Postfach 1780
Telefon (06221) 49101 · Telex 04-61723
1000 Berlin 33 · Heidelberger Platz 3
Telefon (0311) 822001 · Telex 01-83319

Springer-Verlag   New York, NY 10010 · 175, Fifth Avenue
New York Inc.   Telefon 673-2660

Localized Molecular Orbitals and Bonding in Inorganic Compounds

Prof. Dr. H. A. Bent*

Department of Chemistry, School of Physical Sciences and Applied Mathematics, North Carolina State University, Raleigh, North Carolina 27607, USA

Contents

I.	Introduction	1
II.	Antisymmetrization and Localized Molecular Orbitals	4
III.	A Correspondence Principle	8
IV.	Covalent Bonding as a Problem in Classical Electrostatics	13
V.	Covalency Limits	19
VI.	Saturation of Secondary Affinity	22
VII.	Lone Pairs	26
VIII.	Further Uses of Localized Electron-Domain Models	29
IX.	Different Domains for Different Spins	36
X.	Summary	41
XI.	References	45

I. Introduction

One learns at an early age that the world is made up of "objects" — things having the property of *mutual spatial exclusion. Two "objects" are never found at the same place at the same time.*

Traditionally the word "object" has been applied to macroscopic bodies. During the past century, beginning with the work of *van der Waals*, particularly, the concept of spatial exclusion has been applied fruitfully, if approximately, to ever smaller entities: first to individual molecules and ions, and then to their component parts, *e.g.* t-butyl groups, methyl groups [1], and even individual C—H bonds, Fig. 1 [2].

---

* Based on a paper presented at a Symposium on the Stereochemistry of Inorganic Compounds sponsored jointly by the Inorganic Chemistry Divisions of the Chemical Institute of Canada and the American Chemical Society, at Banff, Alberta, June 12—14, 1968.

This study was supported in part by a grant from the National Science Foundation.

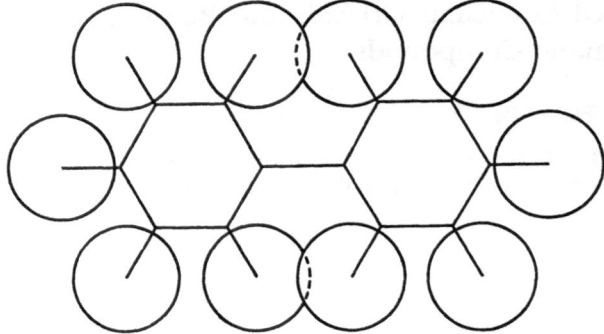

Fig. 1. Model of biphenyl showing van der Waals domains of C—H bonds

Fig. 1 is a scale drawing of the planar configuration of a molecular model of biphenyl using 1.0 Å for the van der Waals interference radius of hydrogen. Analogously, Fig. 2 is a drawing of a *Fieser* [3]- (or, equally well, a *Dreiding* [4]- or Prentice-Hall [5]-) type valence-stick model of cyclohexane in which *spheres* of appropriate size (ca. 2" diameter for Fieser models) have been placed on the model's projecting polar and equatorial sites to represent the van der Waals radii of the electron pairs of the carbon-hydrogen bonds [6]. Fig. 3 is a drawing of the corresponding model of methane showing the van der Waals radii of the electron-domains of methane's valence-shell electrons superimposed on the molecule's conventional graphic formula.

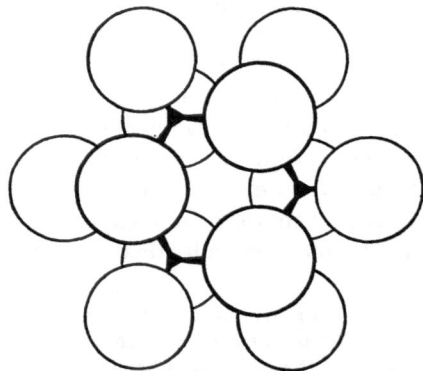

Fig. 2. Modified Fieser model of cyclohexane showing van der Waals domains of of C—H bonds

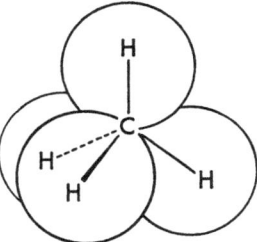

Fig. 3. Graphic formula and electron-domain model of methane. Not shown is the relatively small electron-domain of the carbon atom's $1s$ electrons

It is a small step from *van der Waals*, electron-domain models of the C—H bonds of, *e.g.*, biphenyl, cyclohexane, or methane (Figs. 1—3), to molecular models in which to a first, and useful, approximation *each* valence-shell electron-pair is represented by a spherical, van der Waals-like domain [7]. (Non-spherical domains may be useful for describing, *e.g.*, lone pairs about atoms with large atomic cores, *d*-electrons, and the electron-pairs of multiple bonds; *vide infra*.)

Fig. 4 is a drawing of an all-valence-shell-electron-domain model of ethane superimposed on the molecule's conventional graphic formula. Not shown are the electron-domains of the carbon atoms' $1s$ electrons. In Fig. 4, each valence-stroke, *i.e.* each valence-shell electron-pair of ethane, protonated ("C—H") or unprotonated ("C—C"), is represented by a van der Waals sphere.

The analogy between electron-domain models of hydrocarbons and the experimental facts — particularly the length, but, also, the low

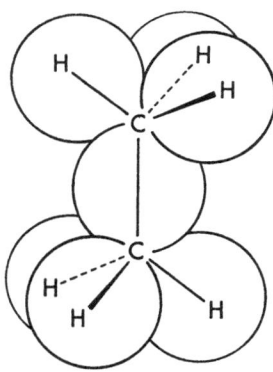

Fig. 4. Graphic formula and equal-sphere-size domain model of ethane

reactivity, of the carbon-carbon bond — is improved if it is supposed that the domain of the electron-pair of a carbon-carbon single bond in *e.g.*, ethane (or diamond or, generally, any unstrained aliphatic compound) has an effective radius significantly smaller (about forty percent smaller) than the electron-domain of the protonated pair of a C—H bond, Fig. 5.

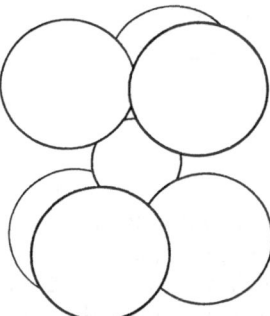

Fig. 5. Improved eletron-domain model of ethane. The domain of the carbon-carbon bond is relatively well-shielded from attack by electrophilic reagents by the domains of the C—H bonds

An analogy exists, also, between electron-pair-domain models and quantum mechanical models of molecules.

II. Antisymmetrization and Localized Molecular Orbitals

Owing to the indistinguishability of electrons, the wavefunction of a molecule's electron-cloud must be antisymmetric in the coordinates of the electrons. Hence, in the orbital-approximation, the wavefunction of a molecule (whose state corresponds to a set of complete electronic shells) can be expressed as a Slater-determinant, each column or row of which is written in terms of a single spin-orbital [8]. As pointed out, however, by *Fock* [9] and *Dirac* [10], and later stressed by *Lennard-Jones* [11] and *Pople* [12], *the orbitals of a Slater-determinant are not uniquely determined, mathematically.*

Adding to or substracting from each other the rows or columns of a determinant does not alter the expanded determinant. *Any orbital-set of a Slater-determinant can be replaced*, therefore, *by any linearly independent combination of the orbitals*, without altering the determinant (except perhaps by a numerical factor). For example, the wavefunction ψ for the four valence-shell electrons of a carbon atom in a ^5S state may be written, in the orbital approximation, either in terms of the $2s$, $2p_x$,

$2p_y$, $2p_z$ orbitals or in terms of the linearly equivalent tetrahedral orbitals te_1, te_2, te_3, te_4 [$te_1 = (^1/_2) (2s + 2p_x + 2p_y + 2p_z)$, etc.] [12]. For all values of the electronic coordinates,

$$\psi = N \, \| \, 2s, \, 2p_x, \, 2p_y, \, 2p_z \, \| = N \, \| \, te_1, \, te_2, \, te_3, \, te_4 \, \|,$$

where $\| \ \|$ indicates a Slater-determinant and N is a normalizing factor.

Before — and, indeed, for sometime after — the development of determinantal wavefunctions, this mathematical equivalence was not widely recognized. Quite early *Pauling* [13] and *Slater* [14] had shown how to create four semi-localized, partially directed, sp^3 orbitals (the tetrahedral orbitals, *te*) by "hybridizing" (*i.e.* by taking linear combinations of) atomic *s* and *p* functions (vide *supra*). Even earlier, in constructing mathematical models of directed bonds, chemists and physicists had intuitively created the p_x and p_y orbitals from the less directional p_{-1} and p_{+1} orbitals of atomic spectroscopy. Still, in the early papers on molecular orbital theory, which were necessarily descriptive and intuitive rather than analytic [15], *Hund* and *Mulliken* generally stopped with the setting up of, and assignment of electrons to, orbitals [16]. The freedom allowed in the selection of these orbitals, owing to the indistinguishability of electrons, was not systematically investigated until 1949. Then, in a series of important papers, *Lennard-Jones* [11] and co-workers [17,18] showed that by "hybridizing" the conventional, symmetry-adapted orbitals of molecular orbital theory, one could often produce a set of mathematically equivalent, semi-localized molecular orbitals closely related to the valence-strokes, lone pairs, and atomic cores of *Lewis's* interpretation, and extension [19], of classical structural theory.

The electron-domains derived from classical structural theory (Figs. 1—5) may be regarded as providing the approximate correlation of electrons of the same spin [20]. Such domains, or loges [21,22], may be called, after *Edmiston* and *Ruedenberg* [23], *localized molecular orbitals* (LMO's). As domains of this type can be generated by maximizing the distance between the centroids of a set of molecular orbitals, or by maximizing the sum of the orbitals' self-repulsion energies, they may be called, also, *exclusive orbitals* [24], or energy-localized orbitals [23]; too, since they are by intention at least qualitatively transferable from molecule to molecule [25], they have been called *molecularly invariant orbitals* (MIO's) [26].

Owing to their transferability and high self-repulsion energies, localized molecular orbitals should be useful in discussions of the correlation problem [27] in quantum mechanics [28]. Indeed, it is found that the correlation energy (the best SCF energy less the observed energy) is remarkably constant for electronic configurations that maintain well-defined localized pairs of electrons with similar relationships among pairs [29] — a result that has led to the view that the chief part of the correlation energy occurs between electrons of opposite spin in the same localized orbital [15].

Recently many authors have discussed procedures for extracting chemically useful information from Hartree-Fock SCF functions through the generation of localized orbitals [30–41]. Rigorously exclusive orbitals, such as those shown diagrammatically in Figs. 1—5 and studied by *Kimball* and co-workers in the early 1950's [7], have generally not been achieved. Significantly, perhaps, the wavefunctions that have been extensively localized have generally been quite approximate, and the results have generally been limited by the constraint that the transformation from one set of orbitals to the other be orthogonal [35]. Different localization procedures generally yield different but similar numerical results [35].

Construction of molecular wavefunctions from "Lewis basis sets" of spherical Gaussian's has been investigated by *Frost* [42] and co-workers [43]. For ethane, *e.g.*, one Gaussian (at least) would be centered in each of the domains of Fig. 4 or 5.

A localized molecular orbital representation is the closest approach that can be achieved, for a given determinantal wavefunction, to an electrostatic model of a molecule [44]. With truly exclusive orbitals, electron domains interact with each other through purely classical Coulombic forces; and the wavefunction reduces, for all values of the electronic coordinates, to a single term, a simple Hartree product.

Linear transformations from delocalized to localized molecular orbitals [and the corresponding inverse transformations [45]] serve, thus, as a bridge connecting the results of quantum mechanical studies with those of classical structural theory. Schematic diagrams of the localized orbitals obtained by *Edmiston* and *Ruedenberg* for the fluorine and nitrogen molecules are shown in Fig. 6 [44]. More detailed drawings of the lone pair and bonding pair orbitals of N_2 are shown in Figs. 7 and 8.

a   b

Fig. 6a and b. Schematic representation of localized molecular orbitals for (a) F_2 and (b) N_2, after *Edmiston* and *Ruedenberg* [44]

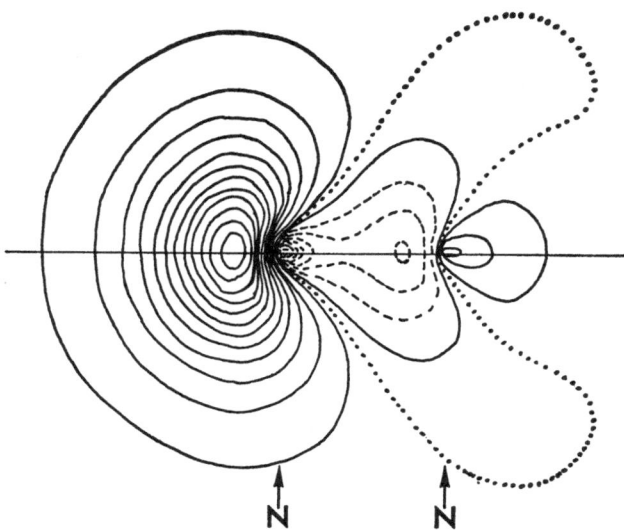

Fig. 7. Contour map by *Ruedenberg* and *Salmon* of the sigma lone pair on N in N_2. The dotted line denotes a node, full lines positive values, dashed lines negative values. The lowest contour represents an absolute value of 0.025 Bohr$^{-3/2}$. The increment between contours is 0.05 Bohr$^{-3/2}$. Courtesy of Professor *Klaus Ruedenberg*. (Note added in proof: The first contour within the dotted contour has been incorrectly copied from Professor *Ruedenberg's* fig. It should be a dashed, not a solid, line.)

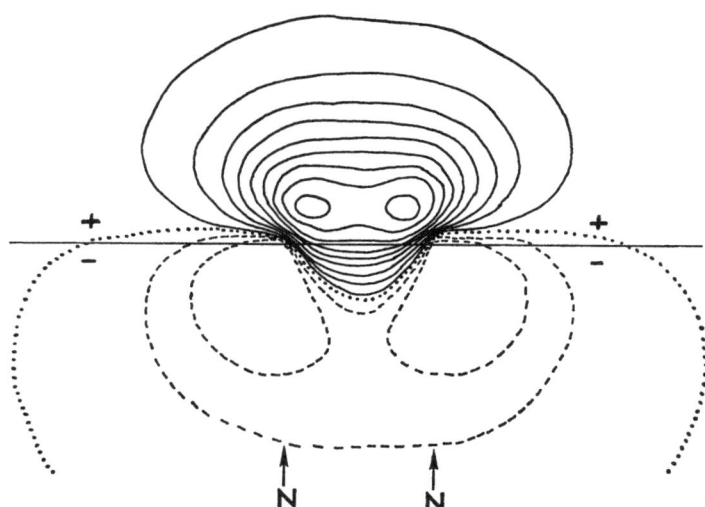

Fig. 8. Contour map of a trigonal bonding orbital in N_2. Courtesy of Professor *Klaus Ruedenberg*

Recently *Gamba* has emphasized [46] that, owing to the antisymmetrization requirement, even for atoms, "A much better *picture* is obtained [over that given by filling successively $1s, 2s, 2p, etc.$ orbitals] by choosing appropriate combinations of the original single particle states so that they have minimum spatial overlap and by filling successively these 'bunched orbitals' with electrons. *The picture that thus emerges* (emphasis added) *is much nearer to the model of close-packed spheres than to any other model.*"

III. A Correspondence Principle

A striking analogy exists between localized molecular orbital, electron-domain models of organic and other covalently bonded molecules (Figs. 3—8) and ion-packing models of inorganic compounds [47].

Fig. 9, from *Warren's* 1929 paper on the crystal structure and chemical composition of amphiboles [48], is a physical picture of the silicon-oxygen chain in diopside. Large circles represent van der Waals-like domains of oxide ions ($r = 1.40$ Å [2]); smaller, dashed circles represent van der Waals-like domains of silicon cations ($r = 0.41$ Å [2]).

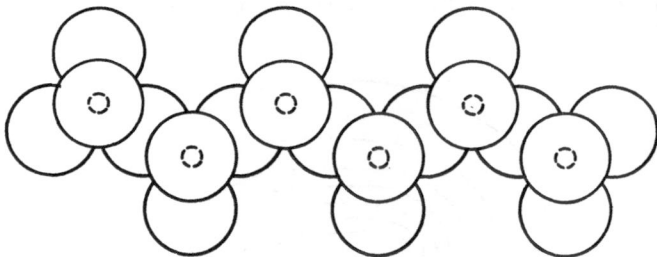

Fig. 9. The silicon-oxygen chain in diopside [48]. Or, a localized molecular orbital representation of a hydrocarbon (see text)

Fig. 9 may be viewed, also, as a localized molecular orbital representation of, *e.g.*, a hydrocarbon (cf. Fig. 13, ref. 7). Thus, replacement of (i) the domains of the Si^{4+} cations (the atomic cores of silicon atoms) by the domains of C^{4+} cations (the atomic cores of carbon atoms; $r = 0.15$ Å [2]), (ii) the domains of the bridging (*i.e.*, bonding) oxide ions by the domains of the electron-pairs of aliphatic carbon-carbon single bonds ($r \sim 0.6_6$ Å [49]), and (iii) the domains of the non-bridging oxide ions by the domains of the protonated electron-pairs of carbon-hydrogen bonds (polarized hydride ions; $r \sim 1.0$ Å, when coordinated by C^{4+} cations), yields a model, analogous to that shown for ethane in Fig. 5, of the all-*trans*, all-eclipsed configuration of *n*-hexane. Briefly stated, a one-to-one correspondence exists between the components of the ionic model of the silicate chain in diopside and the covalent model of a hydrocarbon.

These results may be generalized in a Principle of Correspondence: *There exists an isomorphism between the ion-packing model of heteropolar compounds and the electron-pair model of homopolar compounds.* This correspondence is summarized in Table 1.

Table 1. *An isomorphism* [47)]

Heteropolar compounds	Homopolar compounds
Cations	*Atomic cores*
Relatively large atomic cores. Sizes and shapes approximately independent of chemical environment.	Chemically invariant parts of atoms, nearly. Relatively small, highly charged cations.
Anions	*"Electride ions"*
Negatively charged bodies. Generally larger and more polarizable than cations.	Valence-shell electron-pairs. Charge -2 if unprotonated, -1 if protonated. Generally larger and more polarizable than atomic cores.
Ionic radius	*Electride ion's radius*
An approximate characterization of an ion's domain of influence.	An approximate characterization of an electron-pair's domain of influence.
Crystal	*Molecule*
Large, periodic lattice of cations and anions.	Small, aperiodic lattice of atomic cores and electride ions.
Isomorphic crystals	*Isoelectronic molecules* [50)]
Close analogy in chemical formula and structure. Relative numbers of cations and anions the same.	Identical generalized Ramsey-Lewis-Fajans formulas [47)]. Numbers of atomic cores and electride ions the same.
Basic oxide	*Reducing agent*
Oxide ion donor. Relatively large cations.	Electride ion donor. Relatively large atomic cores.
Acidic oxide	*Oxidizing agent*
Oxide ion acceptor. Relatively small cations.	Electride ion acceptor. Relatively small atomic cores.

Table 1 (continued)

Heteropolar compounds	Homopolar compounds
Coordination compound Close confederation of cations and anions.	*Covalent compound* Close confederation of atomic cores and electride ions.
Coordination site A region of space about a cation that may be occupied by an anion.	*Localized molecular orbital* A region of space about an atomic core that may be occupoed by an electride ion.
Pauling's first rule [51] Each cation is surrounded by a number of anions.	*The Couper-Crum Brown convention* [52] Each chemical symbol is surrounded by a number of valence strokes. *I.e.* (after *Lewis*), each atomic core is surrounded by a number of electride ions.
Coordination polyhedron A description of a cation's anionic environment.	*Sextet, octet, ...* A description of an atomic core's electronic environment.
First coordination shell Spherical sheath about a cation. Generally well-occupied by anions.	*Valence shell* Spherical sheath about an atomic core. Generally well-occupied by electride ions.
Second coordination shell Region immediately beyond the bumps and hollows produced by the anion's in a cation's first coordination shell.	*Outer d-orbitals and antibonding orbitals* Potential energy "pockets" [53] about the electride ions in an atomic core's valence shell.
Coordinatively saturated Coordination shells well-occupied by anions.	*Valence rules satisfied* Valence shells well-occupied by electride ions.
Anion deficient Structure with insufficient anions to complete *separate* coordination polyhedra about *each* cation.	*Electron deficient* Structure with insufficient electride ions to complete *separate* octets about *each* atomic core.
Shared corner Two coordination polyhedra sharing a single anion.	*Single bond* Two octets sharing a single electride ion.
Shared edge Two coordination polyhedra sharing two anions.	*Double bond* Two octets sharing two electride ions.

Table 1 (continued)

Heteropolar compounds	Homopolar compounds
Shared face Two coordination polyhedra sharing three anions.	*Triple bond* Two octets sharing three electride ions.
Effects of multiple sharing on cation-cation distances The more anions two cations share with each other, the less the distance between the two cations.	*Bond orders and bond lengths* Triple bonds are shorter than double bonds, which are shorter than the corresponding single bonds.
Pauling's third rule Shared edges and particularly shared faces destabilize a structure, owing to the cation-cation Coulomb term.	*Baeyer's strain energy* Double bonds and particularly triple bonds destabilize a structure, owing to the kernel-kernel Coulomb term.*
Pauling's fourth rule Cation's with large charges tend not to share anions with each other, owing to the cation-cation Coulomb term.	*Large core-charge strain energy* Atomic cores with large charges tend not to share electride ions with each other, owing to the kernel-kernel Coulomb term.
Irregular polyhedra Coordination polyhedra whose anions are not all shared alike will generally be distorted.	*Non-ideal valence angles* Octets whose electrons are not all shared alike will generally be distorted.
Pauling's fifth rule Mutual repulsions between two cations that share edges or faces of their coordination polyhedra with each other may displace the cations away from the centers of their polyhedra, with obvious effects on internuclear distances and angles.	*Effects of multiple bonds on molecular geometry* Mutual repulsions between two atomic cores that share two or three electrides ions with each other may displace the cores away from the centers of their octets, with obvious effects on bond angles and lengths [52].
Simple bridging ion An anion shared by two cations.	*Ordinary bonding pair* An electride ion shared by two atomic cores.
Non-bridging ion An anion in the coordination shell of only one cation.	*Lone pair* An electride ion in the valence shell of only one atomic core.

* Kernel = Atomic core.

Table 1 (continued)

Heteropolar compounds	Homopolar compounds
Multiply-bridging ion An anion shared by three or more cations.	*Multicenter bond* An electride ion shared by three or more atomic cores.
Transferability of ionic domains Unshared anions occupy about the same space in a cation's coordination shell as do chemically identical shared anions.	*The Lewis [54]-Sidgwick-Powell [55]-Gillespie-Nyholm [56] rule* Lone pairs may often be treated like bonding pairs.
An empirical rule Unshared anions occupy more space in a cation's coordination shell than do chemically identical shared anions.	*Gillespie's rule [57]* Unshared electride ions occupy more space in an atom's valence shell than do shared electride ions.
The Bragg-West rule [58] The size of an anion appear to be larger the larger the smallest cation to which it is coordinated.	*An empirical rule [49]* The size of an electride ion appears to be larger the larger the smallest atomic core to which it is coordinated.
Anion lattice The key to the simple description of ionic compounds. Often close-packed.	*Bond diagram* The key to the simple description of covalent compounds. Often a fragment of a close-packed array [7].
Occupied tetrahedral interstice Cations surrounded tetrahedrally by four anions.	*Couper-Kekule-Van't Hoff-Lewis rules satisfied* Atomic core surrounded tetrahedrally by four electride ions.
Occupied octahedral interstice Cation surrounded octahedrally by six anions.	*Expanded octet* Atomic core surrounded trigonal-bipyramidally, or octahedrally, by 5, or 6, electride ions, *etc.*
Pauling's principle of local electrical neutrality Charges are neutralized locally in crystals.	*Lewis's principle of zero formal charges* Charges are neutralized locally in molecules.

IV. Covalent Bonding as a Problem in Classical Electrostatics

The chief content of the isomorphism displayed in Table 1 is embodied in the phrase "electride ion". By introducing at the outset in the electronic interpretation of chemistry the wave-like character of electrons *and* the Exclusion Principle through the concept of van der Waals-like electron-domains or "electride ions", whose sizes indicate the magnitudes of the electrons' kinetic energies, whose impenetrability [a] simulates, at least approximately, the operation of the Exclusion Principle, and whose charges yield within the framework of the model easily foreseeable effects, one transforms the complex treatment of the covalent bond in quantum mechanics into a simpler, if less precise, exercise in classical electrostatics.

Coulomb's law by itself does not lead, of course, to directions of preferred attraction. Owing, however, to the assumed impenetrability and semirigidity of the charged domains [owing, in other words, to an assumed transferability of the domains' "sizes and shapes" (evidently often approximately spherical)], interactions among positively charged atomic cores and the larger, negatively charged electride ions yield structures of the type illustrated in Figs. 3—5. These structures stand in close analogy with classical models of ionic and covalent bonds (Table 1 and Figs. 1—5) and with the results of recent research on the localizability of molecular orbitals (Figs. 6 and 7). They permit one to see easily, if with not complete certainty, how the chief characteristics of chemical affinity — its saturation and directional character — arise from the joint action of the wave-like character of electrons, the Exclusion Principle, and Coulomb's law.

By exhibiting clearly the basic fact that electrons are wave-like fermions (de Broglie particles that obey the Pauli Exclusion Principle), the LMO-electride ion model of electronic structure enables one to utilize systematically many features of classical physics in developing an "understanding", or "explanation", of the properties of quantum mechanical systems.

Below are five illustrative examples of the explanatory power of classical physics in structural chemistry. In these examples, classical electrostatic interactions are used with the electron-domain representation of molecules to "explain" or to derive: "The New Walsh Rules", the Langmuir-Pauling and Hendricks-Latimer Occupancy Rule, the s-character Rule, the Methyl Group — Tilt Rule, and the Octet Rule.

1. *The New Walsh Rules.* Rules such as the rule that covalent molecules with three heavy atoms ("heavy atom" = any atom other than hydrogen) and 18 valence-shell electrons are bent, while with 16 valence-

---

[a] Not "hardness." It is useful to suppose that the electron-domains are impenetrable but deformable; *vide infra*.

shell electrons the molecules are linear, in their ground states, have been called "The New Walsh Rules" [59]. These rules follow immediately from simple electrostatic considerations.

The arrangement in space of 3 atomic cores and 9 (or 8) electron-pairs that has the lowest (most negative) coulombic energy is shown schematically in Fig. 10. Large circles represent valence-shell electron-pairs; smaller circles labelled A, B, C represent the kernels of the three heavy atoms. For convenience, Fig. 10 has been drawn to show the approximate [b] solution for the problem in two dimensions.

In this instance, the same qualitative results are obtained in three dimensions, where the arrangement of anions about each cation would be a tetrahedral arrangement rather then the square-planar arrangement shown in Fig. 10.

Beneath each tangent-circle drawing in Fig. 10 is shown the corresponding graphic formula. Shared electron-pairs are represented by va-

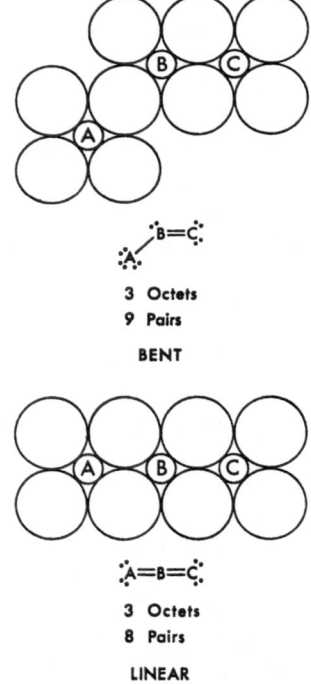

3 Octets
9 Pairs

BENT

3 Octets
8 Pairs

LINEAR

Fig. 10. Tangent-circle, two-dimensional electron-domain models illustrating "Walsh's New Rules"

b) The regularity of the anion-lattice in the top structure of Fig. 10 does not quite correspond to a minimum in the system's potential energy.

lence-strokes, unshared pairs by pairs of dots. The alignment of the heavy-atom skeleton (Is it linear or bent?) is perhaps most easily determined by computing the number of electron-pairs shared between the octets [60]:

(no. pairs shared between octets) $= 4$ (no. octets) $-$ (no. pairs).

2. *The Occupancy Rule.* Having determined the probable articulation of a molecule's electron cloud, one is faced with the question, How should the cations be assigned to the interstices in the anion-lattice? What arrangement of cations with, *e.g.*, charges $+5$, $+6$, $+7$ minimizes the energy of the uppermost structure in Fig. 10? (What, in other words, is the probable structure of nitrosyl fluoride, NOF?) This question was discussed briefly, and informally, by *Langmuir*, in 1921 [61], and, later, by *Pauling* and *Hendricks* [62] and *Latimer* [63].

Three types of interactions contribute to the lattice energies of the (small, aperiodic) "crystals" depicted in Fig. 10: kernel — electron-pair attractions (the only interactions that favor ion-aggregation); kernel-kernel repulsions; and electron-pair — electron-pair repulsions.

The most important interactions, energetically, are those between the cations (the atomic cores) and their *nearest* neighbors (their valence-shell electrons). In absolute magnitude, the sum of the kernel-nearest neighbor interactions for the uppermost structure in Fig. 10 is three-hundred percent larger than the sum of the remaining kernel-electron-pair interactions and, alone, is over twenty percent larger than the sum of all the repulsion terms. Clearly, in minimizing the coulombic energy of such a system, it is of first importance to fill with anions the valence-shells of each cation (*Pauling*'s First Rule).

After the kernel-nearest neighbor interactions, the most important interactions, energetically, are the remaining kernel — electron-pair interactions. The magnitude of their sum for the uppermost structure in Fig. 10 is slightly greater than the sum of the electron-pair — electron-pair repulsion energies, which in turn is larger than the sum of the kernel-kernel repulsion energies. From this one might suppose, erroneously, that the structure's total energy would be a minimum for that arrangement that maximizes the magnitude of the interaction energy of the kernels with the electron cloud; *i.e.*, for the arrangement that places that kernel with the largest charge near the center of the electron cloud; namely, $(+5) - (+7) = (+6)$ (N—F=O).

In fact, although the sum of the kernel-kernel repulsion energies is the smallest of the several sums mentioned (owing to the relatively high concentration of charge within the kernel's domains and the fact that repulsion energies among charges *within the same domain* are not being computed), *changes* in the kernel-kernel repulsion sum with changes in the assignments of the kernels to the interstices of the anion-lattice are approximately twice as large as the accompanying changes in the kernel — electron-pair sum. This is because location of the kernels on the inside of the molecule and the electron-pairs on the outside means that, on the whole, the kernels are closer to each other than they are to electron-pairs outside their own valence-shells. Thus, for a given arrangement of anions, the total energy is generally a minimum for that arrangement that minimizes the cation-cation coulomb term [63]; *i.e.*, for the arrangement that places the kernels with the largest charges in those interstices that share the fewest elements with other interstices; *i.e.*, in the present instance, for the arrangement $(+7) - (+5) = (+6)$ $(F - N = O)$.

3. *The s-Character Rule.* The rule that the s-character of an atom tends to concentrate in orbitals the atom uses in bonds toward electropositive substituents [64,65] follows qualitatively from simple electrostatic considerations, Fig. 11.

Fig. 11 is a drawing of a two-dimensional analogue of the electron-domain model of ethane. Large circles represent valence-shell electron-domains (superimposed on them are the valence strokes of classical structural theory). Plus signs represent protons of the "C—H" bonds. The nuclei of the two "carbon" atoms are represented by small dots in the trigonal interstices of the electron-pair lattice. While these nuclei would not necessarily be in the centers of their interstices, exactly, it can be asserted that an (alchemical) insertion of the two protons on the

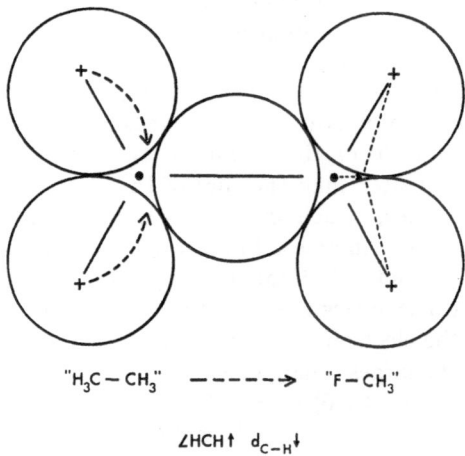

"$H_3C - CH_3$"   -----⟶   "$F - CH_3$"

∠HCH↑ d_{C-H}↓

Fig. 11. Tangent-circle, two-dimensional electron-domain model of ethane and methyl fluoride illustrating the rule that the s-character of an atom tends to concentrate in orbitals the atom uses toward electropositive substituents [64]

left into the adjacent heavy-atom's nucleus (dashed arrows, — which would convert that atom to the two-dimensional analogue of fluorine and the molecule as a whole to the two-dimensional analogue of methyl fluoride — would cause the nucleus on the right to move further to the right (small dashed arrow), thereby increasing the "HCH" angle and decreasing the "H—C" bond lengths (dashed lines), in agreement with the s-character rule, which states that in methyl fluoride the carbon atom would tend to concentrate its s-character in those orbitals it uses in bonds toward the hydrogen atoms. *Hybridization* in this view *is a response of an atomic core to the field of neighboring atomic cores in the molecule* [66].

4. *Tilted Methyl Groups.* Insertion of only one, not both, protons of a "CH$_2$" group into the nucleus of the adjacent heavy atom in the two-dimensional analogue of ethane, Fig. 11, top structure, produces the two-dimensional analogue of the isoelectronic molecule methyl amine, Fig. 11, middle structure. Owing, however, to the dominating attraction of the heavy-atom kernel on the right for the lone pair of the "NH$_2$" group, the symmetrical arrangement of electron-pairs shown in the middle structure of Fig. 11 does not correspond to a minimum in the potential energy surface of the model. A configuration of slightly lower energy is obtained by sliding the "NH$_2$" group over the surface of the carbon-nitrogen bond [67], as indicated in exaggerated fashion in the bottom structure of Fig. 11. This last structure agrees qualitatively with the experimental observation that the methyl groups in, *e.g.*, methyl amine [68], methyl alcohol [69], dimethyl ether [70], and dimethyl sulfide [71] are tilted 2.5—3.5° *toward* the lone pairs on the adjacent atoms.

5. *The Octet Rule.* To the isomorphism exhibited in Table 1 may be added the radius-ratio rules. The statement in crystal chemistry that "rattling is bad", that the number of anions about a cation should not be so great as to create a cavity larger than the cation [72], corresponds to the statement in covalent chemistry that the number of electron-pairs about an atomic core should not be so great as to exceed the number of low-lying, available, valence-shell orbitals.

This correspondence may be expressed in another way. Generally, as one approaches the top of a potential barrier, the density of a system's allowed quantum mechanical levels increases. Thus, to say that an atom has many low-lying, closely spaced electronic levels available for bond formation corresponds to saying that the atom's kernel has a large effective radius; while to say that an atom has few low-lying levels available for bond formation corresponds to saying that the atom's kernel has a small effective radius. *Rules regarding the availability of atomic orbitals for bond formation correspond in structural chemistry to a set of geometrical parameters, the radii of the atomic cores, Fig. 13.*

H. A. Bent

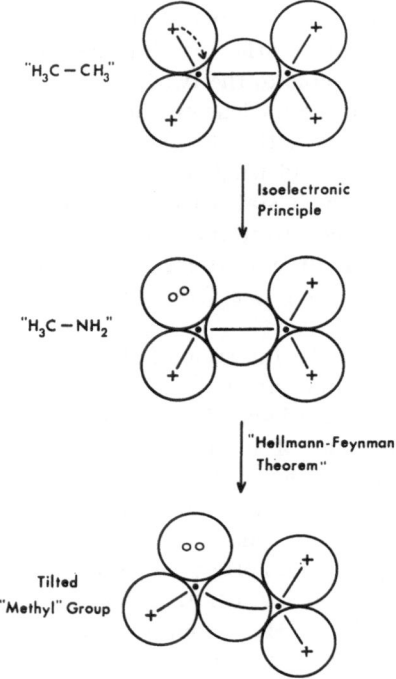

"H₃C – CH₃"

Isoelectronic
Principle

"H₃C – NH₂"

"Hellmann-Feynman
Theorem"

Tilted
"Methyl" Group

Fig. 12. Tangent-circle models for the discussion of tilted methyl groups (see text)

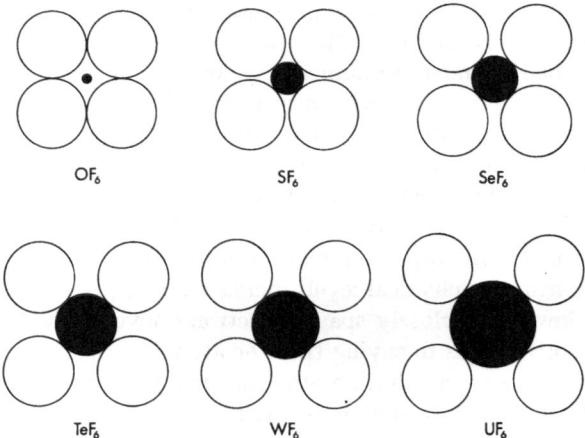

OF₆   SF₆   SeF₆

TeF₆   WF₆   UF₆

Fig. 13. Tangent-circle representation of the local electronic environments about
the kernels of some Group VI hexafluorides

Fig. 13 is a drawing of electron-domain models of some Group VI hexafluorides. Open circles represent the electron-pairs of four of the six bonds to fluorine atoms in a Lewis, single-bond formulation of these molecules. Solid circles represent the atomic cores of oxygen, sulfur, selenium, tellurium, tungsten, and uranium (core radii, in hundreths of Å, 9, 29, 42, 56, 62, and 80 [2]), respectively). These hexafluorides are, in order, non-existent, extra-ordinarily unreactive, hydrolyzed slowly, hydrolyzed completely at room temperature in 24 hours, hydrolyzed readily, and hydrolyzed very rapidly.

The great kinetic stability of sulfur hexafluoride [73,74)], like that of carbon tetrafluoride [73)], particularly toward nucleophilic reagents, may be viewed as arising from the presence about the central atom's kernel (and about the kernels of the fluorine atoms) of a nearly complete, protective sheath of electrons with no "pockets" [52)] of sufficient depth (orbitals of sufficiently low energy) to permit effective coordination with the unshared electrons of an entering nucleophile. The possibility remains, however, of attack by electrophilic reagents, *e.g.*, by strong Lewis acids, such as sulfur trioxide [74)].

The non-existence of oxygen hexafluoride has a simple electrostatic interpretation. Fig. 14 shows the analogous two-dimensional problem. For a relatively small cation [$r(O^{6+}) = 0.09$ Å], coordination by the cation of four anions as shown in structure A is an unstable arrangement. The coordination scheme 2,2 shown in B has a lower energy. Much lower in energy, however, is the coordination scheme 3,1 shown in C.

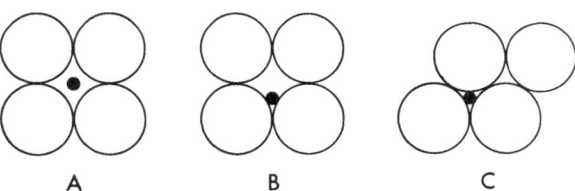

Fig. 14 A—C. A geometrical interpretation of the Octet Rule, in two dimensions (see text). In each figure large, open circles represent domains of an atom's valence-shell electrons. The smaller, solid circle represents the atom's kernel

V. Covalency Limits

The covalency limits of atoms have long been a topic of discussion in valence theory. "Some writers have argued that the increase of the covalency maxima with atomic number is merely a result of the sizes of the atoms," wrote *Sidgwick* in 1933, "and that the reason why, for example, SiF_6^{2-} can exist but not CF_6^{2-} is only that there is not room for six fluorine atoms round the smaller carbon. It is interesting to consider this argument in detail, and see how it has been affected by the increase of our knowledge of atomic dimensions" [75].

Using standard ionic and covalent radii, *Sidgwick* presents his argument graphically, Fig. 15. Each vertical line in Fig. 15 summarizes information on four molecules: MF_6 (octahedral molecules; valency angle 90°) and MF_4 (tetrahedral molecules; valency angle 109.5°), M = C and Si. Outer circles in each group represent halogen atoms. In each instance the uppermost halogen atom makes a valency angle of 90° with the halogen atom on the left of the line and a valency angle of 109.5° with the halogen atom on the right of the line. Halogen atoms are drawn to touch M, "so that the condition of stability is that they should not overlap" [75].

Clearly the ionic model (first column) prohibits too much; it prohibits, *Sidgwick* notes, "even so stable a molecule as carbon tetrafluoride." On the other hand, the covalent model (second column) allows too much; it allows, *Sidgwick* observes, "ample room, not only for the number of attached atoms in actual stable molecules, but for far more than are ever found; even in the imaginary CI_6 there would be space left between the iodine atoms."

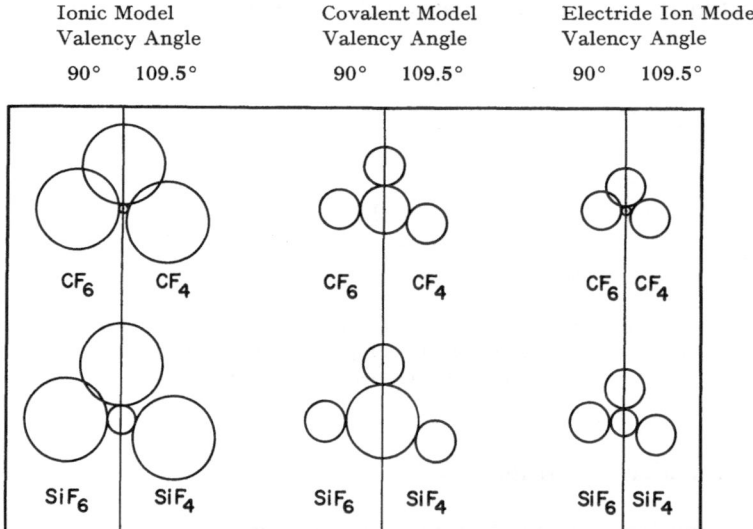

Fig. 15. Tangent-sphere models of CF_4, SiF_4, hypothetical CF_6^{2-}, and SiF_6^{2-}, based on conventional ionic and covalent radii (columns 1 and 2) and the electride ion model (column 3)

"[I]t would seem," concludes *Sidgwick*, "that the steric influence can have nothing to do with covalency limits" [75].

Sidgwick's discussion raises an important question: What are the effective sizes *and shapes* of "atoms" in molecules? From the viewpoint of the electride ion model of electronic structure, *Sidgwick's* circles for the fluoride ions in the first column of Fig. 15 are the wrong shape, if nearly the right *overall* size. In the electride-ion model a fluoride ion is composed of (approximately) spherical domains, but is not itself spherical, in the field of a cation, Fig. 16. Fig. 17 illustrates, correspondingly, the implied suggestion that, on the assumption that non-bonded interactions are not limiting, the covalency limits of an atom will be determined by the radius of the atom's core and by the effective radii, not of the overall van der Waals envelopes of the coordinated ions but, rather, by the radii of the individual, shared electron-pairs.

By an equation given previously [49], the effective radius within molecules of an electron-pair attached to a fluorine atom's core is approximately 0.63 Å, less than half the overall radius of a fluoride ion. Using this smaller value for the radii of the coordinated "anions", revised ionic models may be constructed for the tetrafluoro and hexafluoro complexes of carbon and silicon. Sidgwick-type diagrams for these revised models are given in the third column of Fig. 15. The larger circles represent shared electride ions; small circles represent the atomic cores of the central atoms, as in column 1. The revised tangent-sphere model corresponds better to the known facts than do either of *Sidgwick's* trial models. It is consistent with the existence of CF_4, SiF_4, and SiF_6^{2-} and with the non-existence of CF_6^{2-}.

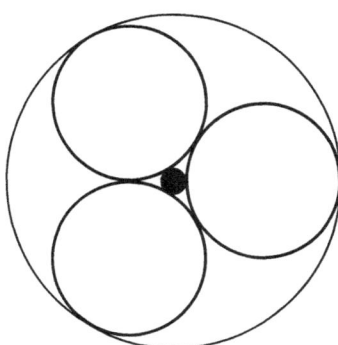

Fig. 16. Two-dimensional, tangent-circle representation of an electron-domain model of a coordinated fluoride ion

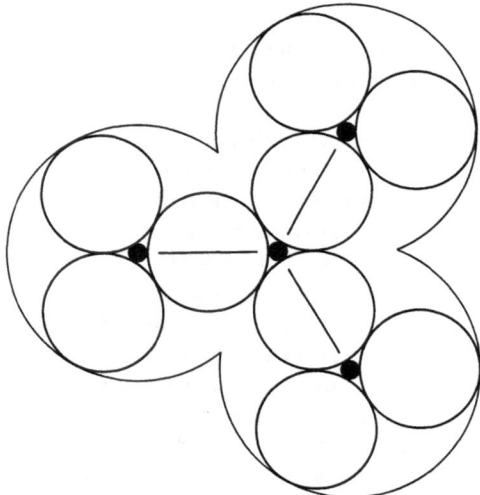

Fig. 17. Two-dimensional, tangent-circle representation of an electron-domain model of CF$_4$. Open circles represent valence-shell electron-domains. Smaller filled circles represent atomic cores. Valence strokes pass through shared electron domains. Note mutual overlap of the ligands' conventional van der Waals domains

After *Werner* [76], and more recently *Rundle* [77], one is led to suggest that *the maximum coordination number of an atom may be considered as having reference to the space around the surface of the kernel of the atom* [78].

VI. Saturation of Secondary Affinity

Neutralization of charge does not cease with the saturation of primary chemical affinity. All stable molecules are surrounded by stray electrostatic lines of force that emerge from not-completely shielded atomic cores (positive patches, hollows, relatively low-lying vacant orbitals, electrophilic centers, acidic sites) and terminate on exposed portions of some electron-cloud (negative patches, bumps, relatively high-lying occupied orbitals, nucleophilic centers, basic sites).

The most exposed portions of molecules' electron-clouds are electron-pairs whose coordinated atomic cores are located largely, or solely, on one side of the pairs, *i.e.*, unshared electrons (particularly those attached to the relatively small kernels of carbon, nitrogen, oxygen, and fluorine atoms and which are, therefore, relatively localized in space; *vide infra*) and, to a progressively lesser extent, the electron-pairs of the bent bonds of "cycloethanes" (olefins) and cyclopropanes, Fig. 18. Lewis acidic sites are usually outward-facing depressions between the valence-shell electron-pairs of medium-sized atoms that have not achieved their covalency maxima. For atoms whose kernels are tetrahedrally coordinated by electron-pairs, these depressions (and, also, the acidic sites off protons in electron-pairs coordinated by highly charged atomic cores) correspond to the lobes of the molecule's lowest-lying, vacant, sigma anti-bonding orbitals.

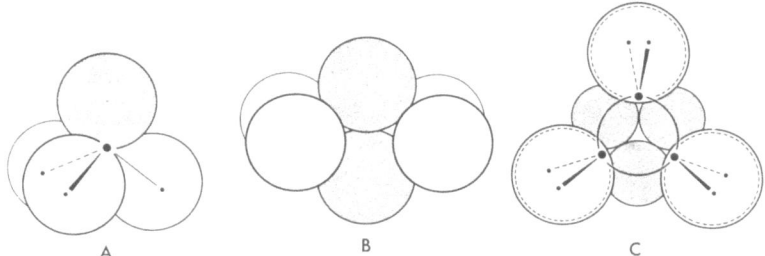

Fig. 18A—C. LMO-electron-domain models of (A) NH_3, (B) C_2H_4, and (C) C_3H_6. Unprotonated domains with sufficient exposure to exhibit nucleophilic activity are shaded

Participation of unshared electron-pairs in the saturation of primary and Lewis-like, secondary acidic sites are compared, schematically, in Figs. 19 and 20. The structures of many molecular complexes formed in the latter fashion have been determined, first and most fully, by *Hassel* and co-workers [79—81].

Fig. 21 is a stylized drawing of a two-dimensional, localized electron-domain model of the molecular complex formed between trimethyl amine (smaller circles, on left) and molecular iodine (larger circles, on right) showing extensive penetration of the conventional van der Waals envelope of the acceptor molecule, I_2, by the protruding electron pair of the donor molecule (van der Waals radius sum for nitrogen and iodine: 3.65 Å; observed $N \cdots I$ distance in the complex $Me_3N \cdot ICl$: 2.30 Å). In this localized electron-domain model of the complex, the immediate electronic environment of the kernel of the acceptor atom may be described as a monocapped tetrahedron (coordination 4;1). The intermolecular interaction producing this configuration has been called a "face-centered bond" [81]. Analogies between the chemical and physical properties of "face-centered bonds" and hydrogen bonds (cf. Fig. 21, lower structure) have been emphasized by several investigators [81,82].

normalH. A. Bent

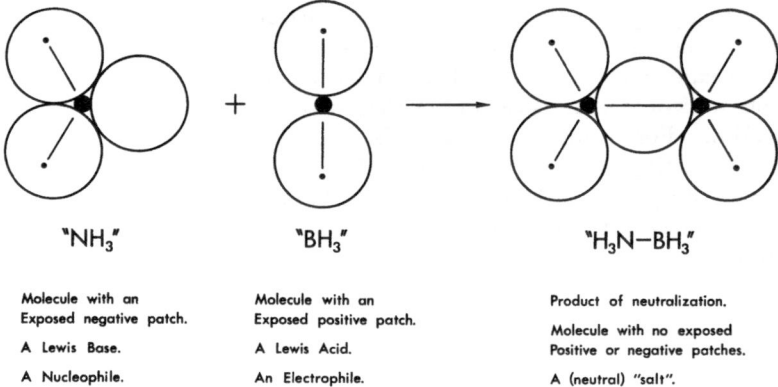

"NH₃"   "BH₃"   "H₃N—BH₃"

Molecule with an Exposed negative patch.	Molecule with an Exposed positive patch.	Product of neutralization. Molecule with no exposed Positive or negative patches.
A Lewis Base.	A Lewis Acid.	
A Nucleophile.	An Electrophile.	A (neutral) "salt".

Fig. 19. Saturation of primary affinity. Two-dimensional representation of an electron-domain model of the formation of a conventional chemical bond: the reaction of a Lewis base (NH₃) with a relatively strong Lewis acid (BH₃).

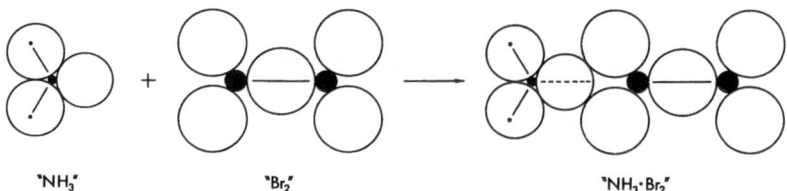

"NH₃"   "Br₂"   "NH₃·Br₂"

Fig. 20. Saturation of secondary affinity. Two-dimensional representation of an electron-domain model of the formation of a molecular complex: the reaction of a Lewis base (NH₃) with a relatively weak Lewis acid (Br₂)

"Hassel-type" complexes (Figs. 20 and 21) have been formed *via* interactions of the acidic sites off "positive halogen" atoms and the unshared electrons in ethers, ketones, sulfides, and selenides. Sometimes both lone pairs of a covalently bonded, bivalent oxygen atom participate in intermolecular interactions. In, for example, the acetone-bromine 1:1 molecular complex, each lone pair of an oxygen atom is shared with a halogen molecule [83]. With very large, sterically unhindered cationic sites, *i.e.*, with metal ions, it appears that both lone pairs of an oxygen atom may be shared with the *same* cation, as shown for the water molecule in B of Fig. 22. The borohydride ion, BH_4^-, with which H_2O is iso-electronic, is known to be coordinated in this fashion by the copper atoms in the borohydridobis(triphenylphosphine)copper(I) complex [84].

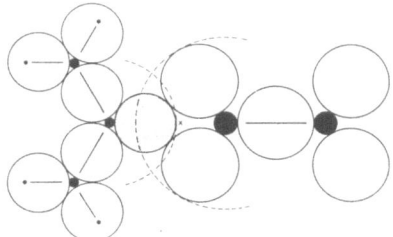

Face-centered Bond

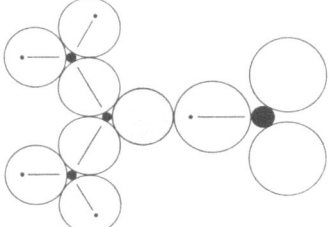

Hydrogen Bond

Fig. 21. Saturation of residual affinity. Schematic, tangent-circle representations of electron-domain models of the molecular complexes $Me_3N \cdot I_2$ and $Me_3N \cdot HI$

And in, *e.g.*, solid ethylmagnesium bromide dietherate, each oxygen atom has, in addition to its two neighboring carbon atoms, a magnesium-atom neighbor that lies almost exactly on the bisector of the COC angle [85].

It would appear that localized molecular orbital, electron-domain models will prove useful in interpretative studies of the structural chemistry of electron-pair donor-acceptor interactions.

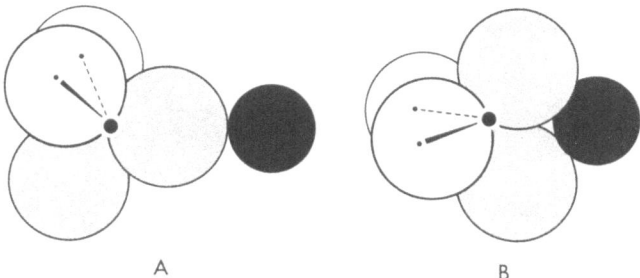

A   B

Fig. 22 A and B. Electron-domain models of a water molecule sharing with a metal cation (solid circle) (A) one and (B) two electron pairs

25

VII. Lone Pairs

A drawing of a two-dimensional, electron-domain model of a conventional Lewis lone pair is shown in Fig. 23. The lone pair and bonding pairs are structurally equivalent; they have identical van der Waals envelopes. Such seems to be nearly the case for lone pairs in the valence-shells of small-core, non-octet-expanding atoms (carbon, nitrogen, oxygen and fluorine).

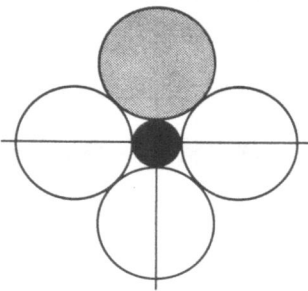

Fig. 23. Two-dimensional representation of an electron-domain model of an angularly-localized, Lewis-type lone pair

With larger atoms, however, a lone pair would not be expected to remain in the angularly localized state shown in Structure A, Fig. 24. In the absence of the other valence-shell pairs, it would envelop the core, becoming an s-type pair. Structure B, Fig. 24, shows schematically the initiation of this process. The lone pair in B is more highly delocalized angularly, if not radially, than a Lewis, bonding-pair-like lone pair. In B the lone pair occupies more space about the surface of the atomic core than does a bonding pair. It may be called a *Gillespie* lone pair.

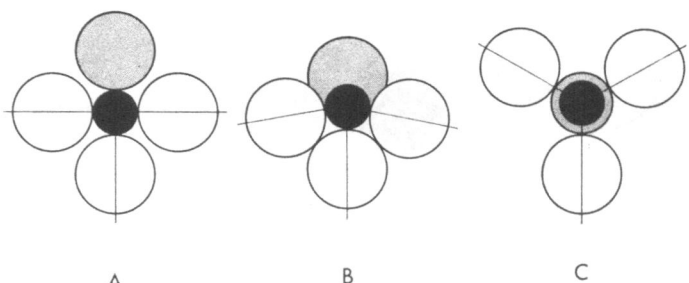

A   B   C

Fig. 24 A—C. Lewis-, Gillespie-, and Sidgwick-type lone pairs

If all the shared electron-pairs about a large atom's kernel were in bonds to exceptionally good leaving groups, a *Gillespie* lone pair would be expected to approach the completely angularly-delocalized state shown schematically in Structure C, Fig. 24. Superficially, Structure C is the structure that would be obtained if one ignored the presence of the lone pair. For this reason the lone pair is often said to be "stereochemically inert". Clearly, however, its presence in the system is structurally significant. Bond angles or coordination numbers, one or the other, are larger, and bond lengths longer, than they would otherwise be. Lone pairs of this type evidently occur in the heavier chalcogenides of lead(II) and tin(II), which have a regular, rocksalt-like structure. They may be called *Sidgwick* lone pairs.

With most ligands, however, a *Sidgwick* lone pair would not be expected to achieve the pure *s*-type configuration shown in Structure C of Fig. 24 or Structure A of Fig. 25. A better representation of a *Sidgwick* lone pair would probably be that shown schematically in Structure B of Fig. 25. Though not yet demonstrated analytically, it is conceivable — indeed, in view of the known facts, perhaps likely — that, by an appropriate assignment of kernel sizes and charges, a *Sidgwick* lone pair could be converted in smooth steps to a *d*-type configuration, such as that shown schematically in Structure C of Fig. 25.

Shapes for lone pairs intermediate between those of Structure C, Fig. 25, and B, Fig. 24, may occur, Fig. 26. Such domain-shapes have, in effect, been used by *Gillespie* [57,86] to account for several features of *Bartell* and *Hansen*'s [87] accurately determined structures of PF_5, CH_3PF_4, and $(CH_3)_2PF_3$, and may be useful, also, in explaining why the axial bond in BrF_5 is evidently shorter than the equatorial bonds.

For some — perhaps infrequently occurring — combinations of kernel sizes, charges, and coordination numbers (*e.g.*, for XeF_6), configurations

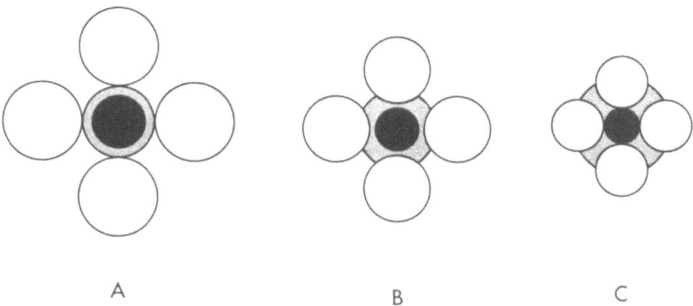

A B C

Fig. 25 A—C. Undeformed (A) and deformed (B and C) Sidgwick lone pairs

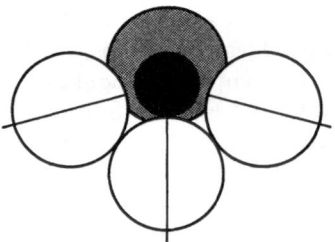

Fig. 26. Gillespie-Sidgwick lone pair

analogous to structures C, Fig. 25, and B, Fig. 24, may differ little in energy, yielding non-stereochemically rigid systems, similar, perhaps, to *Bartell* and *Gavins'* recently proposed model of xenon hexafluoride [88].

There is some evidence in support of the view that an electron-domain's effective volume, if not its shape, is approximately transferable from one system to another. Compress a Sidgwick-type unshared electron on one side and it appears to expand elsewhere, particularly on the opposite (*trans*) side of the kernel, much as one might expect from the form of the kinetic energy operator and the energy minimization principle, which, taken together, require smooth changes in electron density, within a domain.

Square-planar, four-coordinate tellurium(II) complexes, whose structural chemistry has recently been summarized by *Foss* [89], offer illustrations of this highly approximate, constant-volume rule. Fig. 27 is a drawing, approximately to scale, of a schematic, electron-domain representation of a section through the 1:1 complex of benzenetellurenyl chloride with thiourea [90]. Shown are the electron domains of: the tellurium kernel (largest, nearly centrally located solid circle); the ligands' kernels (two chlorine kernels, a sulfur kernel of thiourea, and a carbon kernel of benzene); and the domains of the tellurium atom's shared valence-shell electrons (open circles) and, most schematically of all, the tellurium atom's unshared valence-shell electrons (shaded region).

Figs. 24—26 may be viewed, in succession, as a lone pair of approximate constant volume into, around, and around in which are moved the domains of a central atom's kernel and valence-shell electrons.

The structural chemistry of the tin(II) ion has been reviewed by *Donaldson* [91], who concludes that the effective shape of this ion in crystalline materials may be described as a sphere with a bulge of electron density that prevents the close approach of other bodies along the direction in which the bulge points. "The ns^2 outer electron configuration," he writes, "presents a unique problem in descriptive inorganic chemistry."

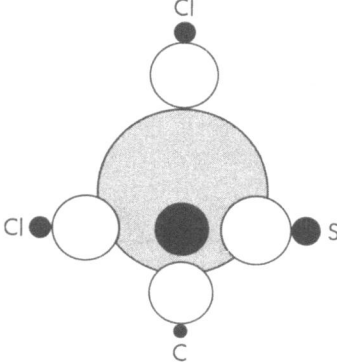

Fig. 27. Schematic representation of the local electronic environment of a tellurium kernel in the 1:1 complex of benzenetellurenyl chloride and thiourea [90]

VIII. Further Uses of Localized Electron-Domain Models

1. *Estimation of Interatomic Distances*. The notion of *transferable inter-ference radii* — that, *e.g.*, a hydrogen atom is approximately the same "size" whether it is attached to a phenyl ring (Fig. 1) or to a cyclohexane ring (Fig. 2); or that a sodium ion is approximately the same "size" whether it is surrounded by chloride ions in NaCl or by, say, bromide ions in NaBr — has found wide application in the estimation of distances between (i) adjacent atoms in adjacent molecules in molecular solids and (ii) adjacent atoms in ionic solids. Extension of these results to the estimation of interatomic distances within covalent molecules through use of the localized electron-domain model and one, new, two-parameter relation (but no new empirical radii) is illustrated in Fig. 28.

In Fig. 28, R is the radius of the electron pair of a linear, one-electron-pair bond, length d; r_a and r_b are *Pauling*'s crystal radii [2]. Empirically, R may be expressed, as indicated, as a constant, approximately equal to the Bohr radius, plus an increment that increases with the size of the *smaller* (and generally the more highly charged) atomic core. It is this atomic core — the one to which in the compound's name the suffix *ide* is added — that largely determines the properties of the shared electron-pair. Some single-bond distances calculated by means of the "R equation" given in Fig. 27 are compared with recently determined interatomic distances in Table 2.

The relationship between atomic radii, ionic radii, van der Waals radii, and the electride-ion model of a shared-electron-pair bond is shown in Fig. 29. M represents a relatively large atomic core of an electroposi-

29

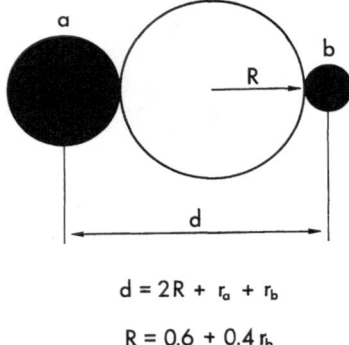

$$d = 2R + r_a + r_b$$

$$R = 0.6 + 0.4\, r_b$$

Fig. 28. Estimation of internuclear distances. Electron-domain model of a partially ionic, linear, one-electron-pair bond. Cf. the bond-charge model of *Parr, R. G.*, and *R. F. Borkman*: J. Chem. Phys. *49*, 1055 (1968) and *Borkman, R. F., G. Simons*, and *R. G. Parr*: J. Chem. Phys. *50*, 58 (1969)

Table 2. *Some calculated and recently observed bond lengths (in Å)*

Compound	Bond	Calc. length	Obs. length
Bis[2-dimethylaminoethyl(methyl) amino-di(methylmagnesium)]	Mg—C	2.12	2.10 [92]
Octamethyldialuminiummonomagnesium	Al—C	1.97	1.97 [93]
Dimethylgermane	Ge—C	2.00	1.95 [94]
Heptamethylbenzene tetrachloroaluminate	Al—Cl	2.17	2.12 [95]
Silicon tetrachloride	Si—Cl	2.09	2.02 [96]
Diboron tetrachloride	B—B	1.76	1.70 [96]

tive element (*i.e.*, a cation). X represents a relatively small atomic core of an electronegative element (*i.e.*, the kernel of an anion). The large, unlabelled sphere represents an electride ion.

Slater has emphasized that the ionic radii for electropositive elements are about 0.85 Å smaller than the atomic radii, while those for the electronegative elements are about 0.85 Å larger than the atomic radii [97]. From Fig. 29 one sees that, to the extent that one may ignore the variation of R with core radii (Fig. 28),

$$(X \text{ or } M\text{'s Atomic Radius}) = (X \text{ or } M\text{'s Kernel Radius}) + R$$
$$(X\text{'s Ionic Radius}) \quad\quad = (X\text{'s Kernel Radius}) + 2\,R.$$

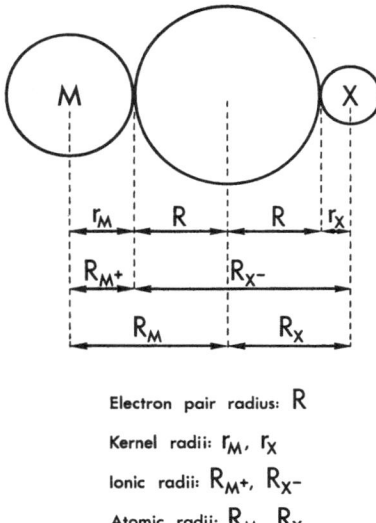

Electron pair radius: R

Kernel radii: r_M, r_X

Ionic radii: R_{M^+}, R_{X^-}

Atomic radii: R_M, R_X

Fig. 29. Atomic and ionic radii. A model for the discussion of the relationships between atomic, ionic, and van der Waals radii.

Hence,

$(X$'s Ionic Radius$) = (X$'s Atomic Radius$) + R$
$(M$'s Ionic Radius$) = (M$'s Kernel Radius$) = (M$'s Atomic Radius$) - R$.

With $R = 0.85$ Å, the equation $R = 0.6 + 0.4\,r$ (Fig. 28) yields for r the value 0.6 Å, close to the mean value of the kernel radii of the more familiar elements.

From Fig. 29, "We can understand", to quote *Slater* [97], "how the sum of ionic radii for an electropositive and electronegative element is able to lead to almost exactly the same result as the sum of the atomic radii of the same elements".

From Fig. 29 one sees, also, that

$$(X\text{'s van der Waals Radius}) = (X\text{'s Kernel Radius}) + 2\,R$$
$$= (X\text{'s Ionic Radius}).$$

Additionally, from Fig. 29 one sees that, if, as proposed by *Frost* [42], a spherical gaussian function is a fair representation of the distribution of charge within an electride ion, *there should be*, as found by *Slater* [97], *"a very good correlation, and in many cases practically an equality, between the atomic radii . . . and the calculated radius of maximum radial charge density in the outermost shell of the atom".*

31

2. *Stereochemistry of Electron-Pair-Coordination-Number 5.* Atoms that contain in their valence shells five electron-pairs present interesting stereochemical problems to which several of the ideas developed above may be applied.

Gillespie has formulated a "90°-Degree Rule" [57c)] that predicts correctly the distribution of lone pairs over axial and equatorial positions in the trigonal-bipyramidal configurations of compounds of the type (E = lone pair) AX_4E (*e.g.* SF_4, $(CH_3)_2TeCl_2$), AX_3E_2 (*e.g.* ClF_3, $C_6H_5ICl_2$), and AX_2E_3 (*e.g.* ICl_2^-, XeF_2). *Gillespie*'s rule works, also, for electron-pair-coordination-number-6 compounds of the type AX_4E_2 (*e.g.* ICl_4^-, XeF_4). The same predictions are obtained from a simple electrostatic model in which a halogen (or other) ligand is represented by a $+1$ point-charge (the effective net charge, seen from a distance, of a halogen atom's $+7$ kernel and 6 unshared electrons) set at twice the distance from the central atom's nucleus as -2 point-charges representing the central atom's valence shell electron-pairs.

In trigonal bipyramidal coordination the axial bonds are longer than the equatorial bonds, for all non-transition elements that have been observed in that coordination [98)]. This stereochemical feature might arise from intrinsic departures of the central atom's core from spherical symmetry [99)], from nonbonded interactions, and/or from repulsions among electrons in the central atom's valence shell. The importance of the latter interactions is suggested by the fact that the difference in lengths of axial and equatorial bonds to identical ligands is (with the exception of fluorine) generally more than or less than 0.15 Å depending on whether there are or are not lone pairs present in the central atom's valence shell. [For fluorine (a poor leaving group), the difference always is less than 0.11 Å.]

The intrinsic tendency of a lone pair to encircle an atomic core, pushing bonding pairs closer to each other (initially) (Figs. 24 A and B) and outward (Figs. 24 C and 26) is further suggested by the structures of the compounds R_2TeX_2E examined by McCullough and co-workers [100–102)] (*R* is an organic group, *X* a halogen, *E* a lone pair). In the direction of the axially located halogen ligands, the apparent size of the tellurium atom — obtained by subtracting the ionic radius of the halogen from the observed $Te-X$ distance — increases in going from $X = Cl$ to $X = I$, *i.e.*, with increasing "goodness" of the leaving group (Cl^-, Br^-, or I^-).

Concentration of the expansive effect of the lone pair upon the bond to a single, good leaving group may so lengthen the bond as to suggest an ionic formulation for the compound. Crystalline trimethylselenonium iodide, *e.g.*, has been described as composed of the ions $(CH_3)_3Se^+$ and I^- [103)]. And trimethyltelluronium bromide is a 1:1 electrolyte in dimethylformamide [104)].

Numerous authors have cited the halide-ion donor properties and tendency toward ionization of the Group VI tetrahalides, particularly SCl_4 [105], $SeCl_4$ [106–107], $TeCl_4$ [106–108], and $TeBr_4$ [107–108], but not the fluorides such as SeF_4 [109]. Toward BF_3, dimethylselenium dihalides, rather than sharing (directly) the unshared pair on selenium, evidently transfer halide ions to the Lewis acid, yielding ionic monohalogendimethylselenium(IV) tetrahaloborates [110].

Generally, for compounds that contain lone pairs attached to atoms of second- and later-row elements, the model of angularly delocalized lone pairs about large-core atoms is useful in rationalizing (i) the low *Brønsted* and *Lewis* bascicity of such electrons; (ii) the abnormal lengths of bonds to adjacent groups, particularly good leaving groups; (iii) the tendency for the central atom to ionize off 'a good leaving group; and perhaps, too, should the lone pair achieve a Sidgwick-type configuration (with a corresponding overall enlargement of the core) in rationalizing (iv) an apparent tendency for a central atom whose coordination number is already relatively large further to increase its coordination number, in condensed phases, through polymerization [111].

Interestingly, the geometrical condition that a spherical atomic core be large enough to touch simultaneously five spherical ligands in trigonal bipyramidal coordination (smallest bond angle $90°$) is the same condition as the non-rattling condition for octahedral coordination. In both instances the atomic core must be large enough to subtend an angle of $90°$ when touching two mutually tangent, spherical ligands. From the viewpoint of the electride ion model, therefore, *an atom that can form compounds in which it has an electron-pair-coordination-number 5 should be able to form compounds in which it has an electron-pair-coordination-number 6*. Such seems to be the case. It would appear that there are no exceptions to this "if-5-then-6" theorem. Conversely, one predicts that in covalent compounds, the *small-core elements carbon, nitrogen, oxygen, and fluorine will never be found in trigonal bipyramidal coordination with two equally distant axial groups* [112].

Closely related to the "if-5-then-6" theorem is the fact, frequently noted, that, for points on a sphere repelling each other according to an inverse square law, the difference in energy between a trigonal bipyramidal arrangement (cf. A in Fig. 30) and a tetragonal pyramidal arrangement (cf. B in Fig. 30) is relatively small. Thus, as suggested by *Berry* [113], interchange of axial and equatorial groups by an intramolecular mechanism (Fig. 30) may be relatively easy. Similar remarks hold for several higher coordination numbers.

3. *Multicenter Bonding.* While the valence stroke of classical structural theory is for many purposes a useful representation of an electron-domain created by the fields of *two* atomic cores (for every line-segment

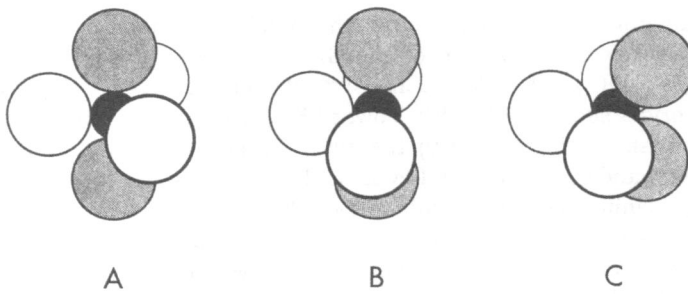

A   B   C

Fig. 30 A—C. The Berry mechanism for electron-pair-coordination-number 5. (A) Trigonal bipyramidal coordination. Ligands bonded through shaded electron-domains are in axial positions. (B) Tetragonal pyramidal coordination. A slight distortion of structure A. (C) Trigonal bipyramidal coordination. A slight distortion of structure B. Ligands bonded through shaded electron-domains are now in equatorial positions

has precisely *two* ends), a segment of a line is not generally used to represent an electron-domain created by the field of a *single* atomic core (a lone-pair domain), nor can it be used — or be conveniently modified — to represent an electron-domain produced by the fields of *three* or more atomic cores. For the latter, a dashed polygon is often used. A three-center bond, *e.g.*, is often represented by a dashed triangle.

Sometimes a three-dimensional model gives a clearer physical picture of a chemical structure than does any stylized, two-dimensional drawing. In conventional ion-packing models of, *e.g.*, BeF_2, MgF_2, CaF_2, NaCl, and CsCl, it is easy to see, if not to show, schematically, in two dimensions, that each anion is shared by — *i.e.* is touching simultaneously — 2,3,4,6, and 8 cations. The Correspondence Principle (Table 1) suggests that *the multiple-shared-spherical-anion model of crystal chemistry might be useful in accounting for the structures of "electride-ion-deficient" covalent compounds.* Indeed, in many instances, localized orbitals, whether 1-center orbitals (for conventional lone pairs), 2-center orbitals (for conventional bonding pairs), or *n*-center orbitals, $n > 2$ (for "multicenter bonding"), can be represented — to a first (and often useful) approximation — by spherical electron-domains.

Spherical-domain models of three-center bonds in localized-molecular-orbital models of a nonclassical carbonium ion, B_4Cl_4, and $Ta_6Cl_{12}^{2+}$ have been described [49,52]. A drawing of a spherical-domain model of the methyl lithium tetramer, $(LiCH_3)_4$, is shown in Fig. 31. Large, outer circles represent domains of electron-pairs of C—H bonds. Solid circles represent domains of Li^+ ions. Shaded circles represent 4-center lithium-lithium-lithium-carbon bonds — *i.e.*, electron-pair domains that touch, simultaneously, three lithium ions and the kernel of a carbon atom. The observed value for the ratio of the lithium-lithium distance in $(LiCH_3)_4$ to the lithium-carbon distance is 1.12. For the model shown in Fig. 31 — with, however, all domains taken to be the same size (*i.e.*, with $r_{CH} = r_{Li} + = r_{LiLiLiC} \sim 0.66$ Å, by the equation in Fig. 28), and with the kernels of the carbon atoms in the centers of their tetrahedral interstices (the most critical — and best — assumption being the one $r_{Li} + = r_{LiLiLiC}$) — the calculated value for the Li-Li/Li-C ratio is 1.14.

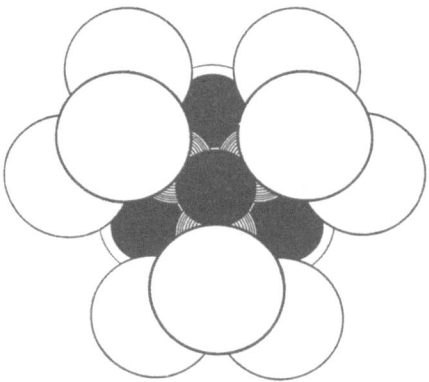

Fig. 31. Electron-domain model of $(LiCH_3)_4$

4. *Metals*. In heteropolar compounds multiple sharing of anions by cations is most pronounced in compounds where the number of anions per cation is relatively small, much smaller than the normal coordination numbers of the cations, *i.e.*, in the halides and chalcogenides of the elements of Group I and II of the Periodic Table. By the Correspondence Principle, one anticipates that models showing multiple sharing of localized *electron-domains* by atomic *cores* will be particularly useful in accounting for the properties of substances for which the number of valence electrons per atomic core is relatively small. Such models for metals from Groups I and II, with electrons — or pairs of electrons — playing the role of conventional anions, have, in fact, been proposed by numerous authors [72,114-120]. These models account directly, if yet only qualitatively, for many of the important chemical and physical properties of metals. While hardly new, these models, with anticipated refinements, may fulfill in other respects *Houston*'s hope "that in due time there may emerge [in the study of metals] some new ideas which are as significant and simple as the treatment of nuclear vibrations by means of normal coordinates, as helpful as the application of Fermi statistics to the electrons, and as pictorial as the treatment of electrons as waves" [121].

IX. Different Domains for Different Spins

Fig. 32 shows on the left a conventional, localized domain model of the electronic environment of an atom that satisfies the Octet Rule. Each domain is occupied by two electrons. It is well known, however, that *the assumption of two electrons per orbital is unnecessarily restrictive* [27,122]. Better energies are obtained in quantum mechanical calculations if different orbitals are used for electrons of different spins, a fact first demonstrated in quantitative calculations on helium by *Hylleraas* [123] and *Eckart* [124]. Later, this "split-orbital" method was applied to π-electron systems [27,125]. Its general application to chemical systems has been developed by *Linnett* [126].

Linnett's procedure may be viewed as a refinement of classical structural theory [99]. In *Linnett's* theory, the *van't Hoff-Lewis* tetrahedral model is applied twice, once to each set of spins, with the assumption that, owing to coulombic repulsions between electrons of opposite spin, there may be, in some instances, a relatively large degree of spatial anticoincidence between a system's two spin-sets.

On the right in Fig. 32 is an electron-domain representation of *Linnett's* model of an Octet-Rule satisfying atom in field-free space. For domains of (i) fixed size and distribution of charge, (ii) fixed distances from the nucleus, and (iii) fixed tetrahedral disposition *with respect to other domains of the same spin-set*, three of the four contributions to the total energies of the two structures in Fig. 32 are identical, namely the energies arising from (i) electronic motion, (ii) nuclear-electron attractions, and (iii) electron-electron repulsions *between electrons of the same*

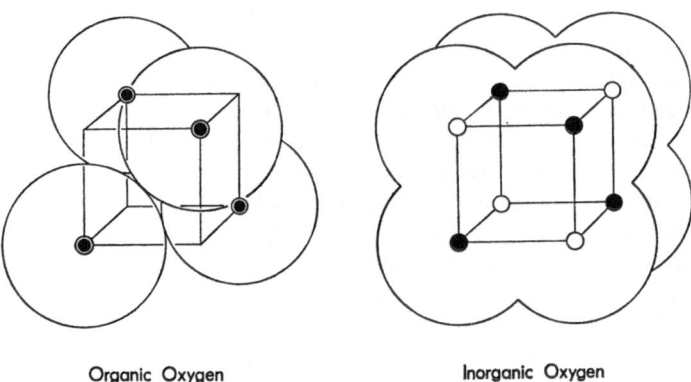

Organic Oxygen   Inorganic Oxygen

Fig. 32. Electron-domain representation of (left) strong-field and (right) weak-field models of an octet, after *Linnett* [126]

spin. Between electrons of *opposite spin*, however, repulsions are less for the structure on the right than for the one on the left. In the absence of external fields, the structure on the right with partially anticoincident spin-sets is presumably a better representation of, *e.g.*, an hypothetical O^{2-} ion than is the structure on the left with fully coincident spin-sets.

The relatively weak fields of two Li^+ ions are evidently not able to force into coincidence the two spin-sets of an O^{2-} ion, for it is reported that in the gas phase the molecule Li_2O is linear [127]. With, however, an increase in charge and decrease in size of the kernels of the ligands, the tendency to produce spatial pairing in the bonding — and, hence, also, in the non-bonding — regions about the central atom increases and the bond angle at oxygen gradually decreases toward the (approximately) tetrahedral angle characteristic of organic (that is to say, small-core) compounds. The data have been reviewed by *Gillespie* [128]. In inorganic compounds average values for the bond angles $Si-O-Si$, $P-O-P$, and $S-O-S$ are, respectively, 137, 128, and 118°. In Cl_2O the bond angle is 110.8°.

In the *Linnett*, weak-field model of an octet, the spin-densities off opposite corners of the octet are of opposite signs, Fig. 32. This suggests that two, identical magnetic ions coordinated colinearly by an oxide ion might be coupled antiferromagnetically, Fig. 33, as, in fact, has been

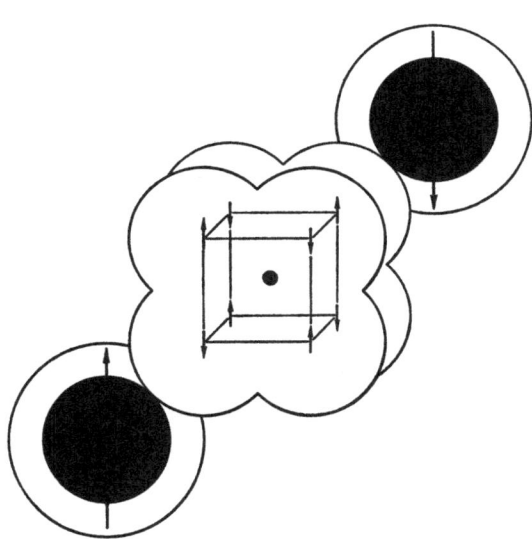

Fig. 33. Antiferromagnetic coupling of two magnetic cations via an intervening anion. Cf. *Halpern, V.*: Proc. Royal Soc. (London) *291*, 113 (1966)

observed, first for MnO [129)], later for several transition metal fluorides with the WO₃ structure [130)], and, also, for such species as $(Cl_5Ru-O-RuCl_5)^{4-}$ [131)] and $(Cl_5Re-O-ReCl_5)^{4-}$ [132)].

One of the earliest applications of the method of different orbitals for different spins appears in *Slater*'s classic study of the cohesion of monovalent metals [133)]. In *Slater*'s model of the body-centered cubic structure of the alkali metals, the lowest unperturbed, zero spin-state is taken to be one in which the valence electrons about the atoms at the cube-corners have one spin, spin α, while those at the cube-centers have the opposite spin, spin β.

The upper drawing in Fig. 34 is a schematic, electron-domain representation of the spin-density in a plane through two neighboring corner atoms and the adjacent center atom in *Slater*'s model of the alkali metals. Solid circles represent the atoms' kernels (M⁺ cations). The Pauli Exclusion Principle permits domains occupied by electrons of opposite spin to overlap (corner atoms with the central atom), but prohibits overlap between domains occupied by electrons of the same spin (corner atoms with corner atoms).

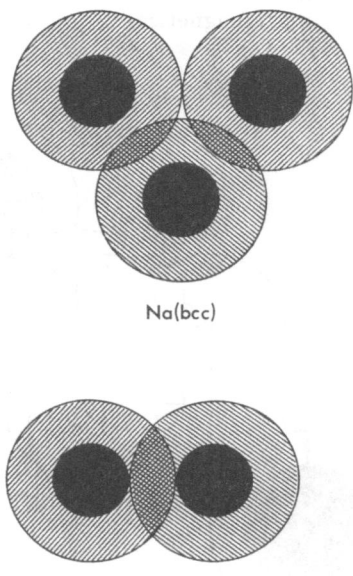

Na(bcc)

Na₂(g)

Fig. 34. Top: Electron-domain representation of Slater's model of the alkali metals [133)]. Bottom: Electron-domain representation of an analogous model for the corresponding gaseous dimers

Since the shortest interatomic distance in the metallic state of the alkali metals, M(bcc), is greater than the interatomic distance in the corresponding gaseous dimer, $M_2(g)$, the length of the body-centered unit cell, a, is shown in Fig. 34 as being fixed by contact between valence-shell domains occupied by electrons of the *same* spin (those at adjacent corners). This model predicts that, insofar as the sizes and shapes of electron-domains are transferable, the distance of closest approach of two isolated alkali atoms should be (bottom drawing, Fig. 34) one-half the unit length, $a/2$, plus the corresponding ionic radius, r_{M+}. This sum is close to the observed interatomic distance for the gaseous dimers of sodium and potassium, Table 3.

Table 3. *A relation between the interatomic distances in solid and gaseous alkali metals (see Fig. 34) (Distances in Angstroms)*

Metal, M	a(bcc)	r_M+ [2)	$a/2 + r_M+$	i. a. d. in $M_2(g)$
Li	3.50	0.60*	2.35	2.67
Na	4.28	0.95	3.09	3.08
K	5.33	1.33	3.99	3.92

* Most authors use a slightly larger value, ca. 0.68 A.

The *core model* of Fig. 34 (atomic cores surrounded — but not penetrated — by valence-shell electrons) has a long history. It was advanced by *Ramsey*, in 1908, who referred to an atom's outer electrons as a "rind" [134]; by *Lorentz*, in 1916, who viewed metals as a crystalline arrangement of hard, impenetrable cations with valence electrons in the interstices [135]; by *Langmuir*, in 1919, who noting (after *Faraday* [136]) and many others) the relatively large molar volume of metallic sodium, postulated a specific repulsion between a completed octet and a single free electron [137]; by *Rice*, in a classic paper on the alkali and alkaline earth metals, in 1933, who, after the fashion of *van der Waals*, took the volume available to a metal's valence electrons as the metal's atomic volume less an "intrinsic ionic volume" which, in anticipation of the theory of pseudopotentials (*vide infra*), was assumed to act "to some extent like a region into which it [a valence electron] cannot go" [138]; by *Stokes*, in 1964, who concluded that in a metal crystal the ions have a covolume equal to "the volume of spheres having the radii found from studies of alkali halide crystals and which is effectively inaccessible to the valency electrons" [139]; by *Raich* and *Good*, in 1965, who treated the conduction electrons as a gas of uniform density in a spherical shell "around impenetrable spheres representing the ion core" [140]; and by *Cutler*, in 1967, who, following *Rice* and later workers, found that the empirically determined effective radius of an ion in a metal is generally comparable with its conventional ionic radius [141].

The potential energy function for a valence electron in the core model is compared with the pseudopotential of *Phillips* and *Kleinman* [142] and *Cohen* and *Heine* [143] in Fig. 35. The potential energy function for the impenetrable core model is given by the dashed line. It is Coulombic down to a core radius R_c, at which point it rises to infinity. According to the cancellation theorem of pseudopentential theory [143-144], the negative potential energy of a valence electron near an atomic nucleus is nearly cancelled by the positive kinetic energy associated with the rapid oscillations of the wave function imposed by the conventional — if not essential [145] — requirement that the wave functions of the valence electron be orthogonal to the wave functions of all the core-electrons. The result is an effective cut-off coulomb potential shown schematically in Fig. 35 by the solid line [146].

To account for the magnetic hyperfine interaction in, *e.g.*, the ground state of a lithium atom, it is necessary in the core model to go beyond the usual, restricted Hartree-Fock approximation, which assumes that the spatial parts of one-electron orbitals are independent of m_s (the spin-

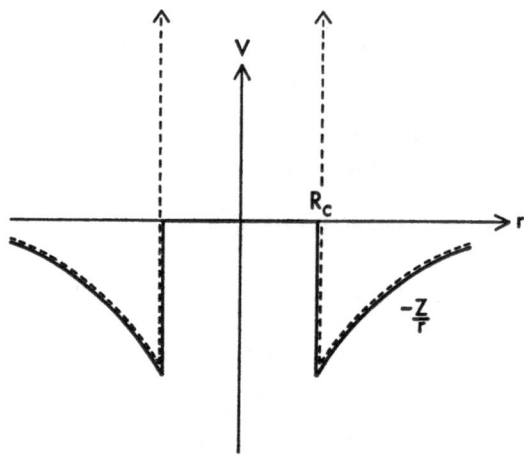

Fig. 35. Dashed line: Potential energy function for a valence electron in the impenetrable core model. R_c = core radius. Solid line: Pseudopotential, after *Austin* and *Heine* [146]

equivalence restriction), and which therefore does not permit core electrons to make a contribution to the magnetic hyperfine interaction [147]. For an atom with an odd number of valence electrons, different radial functions should be used for core electrons with different spins.

The use in the quantum mechanical calculations of discontinuous trial functions constructed from exclusive orbitals has been discussed by *Weare* and *Parr* [148] and *Hall* [149], who establish that, by using, e.g., the *Green*'s function to turn *Schrodinger*'s differential equation into an integral equation [149], one can do *a priori* quantum mechanical calculations on molecules using as basis functions atomic orbitals on different nuclei that are rigorously cut off so that no orbital on one atom has a non-vanishing overlap with any orbital on another atom [148].

X. Summary

The development of useful models of matter prior to 1926 shows that, while a fundamental — *i. e.* widely applicable — explanation of the properties of matter may be, in principle, a quantum mechanical problem, many problems can be treated in a simpler fashion. Even complicated systems have their simpler aspects [150].

Clementi asserts, however, that it is becoming increasingly apparent that any "all-electron" quantum mechanical treatment of a moderately complex molecule is not feasible, even with modern high-speed computers [151]. "Is there," *Whyte* asks, "no short cut from the postulates of physics to our visual observations?" [152].

A significant simplification in the electronic interpretation of chemistry can, in fact, be achieved by introducing the Exclusion Principle at the very beginning of the discussion [153]. The justification for this procedure lies chiefly in the simplicity of the results, their visualizable character, and their explanatory power.

The Exclusion Principle can be expressed in several ways: in *Pauli*'s early statement that no two electrons can have all their quantum numbers the same; or, more satisfactorily (since individual quantum numbers are, strictly speaking, physically meaningless in systems of strongly interacting particles) in *Heisenberg*'s later statement that the wave function for a system of fermions must be antisymmetric.

The antisymmetric requirement is an intrinsic property of electrons, along with their charge, mass, and spin. Any 'state' that violates the antisymmetry requirement is absurd as a state in which an electron has, *e. g.*, a spin of three halves or zero mass [154].

The antisymmetry requirement implies that electrons with the same spin cannot be at the same place at the same time. *Electrons of like spin tend to avoid each other.* "This effect is most powerful," writes *Lennard-Jones* [155], "much more powerful than that of electrostatic forces. It does more to determine the shapes and properties of molecules than any other single factor. It is the exclusion principle which plays the dominant role in chemistry. Its all-pervading influence does not seem hitherto to have been fully realised by chemists, but it is safe to say that ultimately it will be regarded as the most important property to be learned by those concerned with molecular structure."

The Exclusion Principle endows quantum mechanical systems with a property analogous in many respects to the classical concept of impenetrability [156]. This property finds expression in classical structural theory in the concept of molecular, van der Waals domains that may touch and deform one another but do not overlap; in the concept of ionic spheres of influence that, while polarizable and compressible, are effectively impenetrable; and in the well known, if seldom articulated, theorem that the valence strokes of classical structural theory never cross one another [78]. Taken with *Lewis*'s identification of the valence-stroke as *precisely two* electrons, this non-crossing theorem virtually demands (in retrospect) a wave-like character for electrons and an exclusion principle.

The evolution of the valence stroke from a primitive, irreducible element of chemical theory to a construct to which the laws of modern physics may be applied is summarized in Fig. 36.

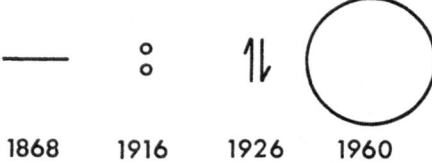

| 1868 | 1916 | 1926 | 1960 |

Fig. 36. A short history of the introduction into structural chemistry of the discoveries of electron-physics

The concept of (approximately) transferable, localized electron-domains provides a link between quantum physics and classical chemical theory and serves to clarify, from the viewpoint of physics, the status of classical chemical concepts. This link provides a chemist, therefore, with an intuitive understanding of quantum mechanical relations, in the sense that it permits one to guess qualitatively, through the use of classical chemical theory, what answers rigorous applications of the quantum mechanical formalism would give when applied to simple chemical problems [157]. Through the Correspondence Principle, the electron-domain representation enables one to utilize, also, in the systematic development of a theory of small-core, covalent compounds nearly every feature of the structural theory of ionic compounds [157]. The chief results obtained in this way are summarized in the following Principles of Structural Chemistry.

Covalent compounds may be viewed as ion-compounds.

Cations = atomic cores (usually relatively small cations).
Anions = valence-shell electrons.
Generally $r_{anion} > r_{cation}$.

A polyhedron of anions, *which may be localized electrons*, is formed about each cation. All compounds may be viewed as coordination compounds.

A cation's coordination number is determined, in part, by the radius ratio. Rattling is bad. Hence,

Small cores $(0 < r < 0.25$ Å$)$ have an EPCN (electron pair coordination number) $= 4$.
Large cores $(r > 0.25$ Å$)$ may have an EPCN > 4.

In anion-deficient structures, polyhedra may share corners, edges, or faces.

Edge- and face-sharing destabilizes a structure, owing to the cation-cation Coulomb term.

Highly charged cations tend not to share polyhedron elements with each other, owing to the cation-cation Coulomb term.

In edge- and face-sharing, and in mutual sharing by highly charged cations, cation-cation repulsions may operate to displace the cations, if small, from the centers of their coordinated polyhedra.

Anions may be shared by more than two cations.

Unshared electrons tend to envelop the atomic cores to which they are coordinated.

The chief differences between small-core, organic compounds and large-core, inorganic compounds are summarized schematically in Fig. 37. At a small-core site (Fig. 37, left drawing), one finds, typically, a coordinatively saturated, octet-rule-satisfying, relatively unreactive electrophilic center whose reactions are kinetically controlled and proceed via back-side attack with inversion of configuration and much bond-breaking preceeding the transition state. At a large-core site (Fig. 37, right drawing), one finds, typically, a coordinatively unsaturated, variable-valence, relatively reactive electrophilic center whose reactions are thermodynamically controlled and proceed often via front-side attack with retention of configuration and much bond-making preceeding the transition state.

H. A. Bent

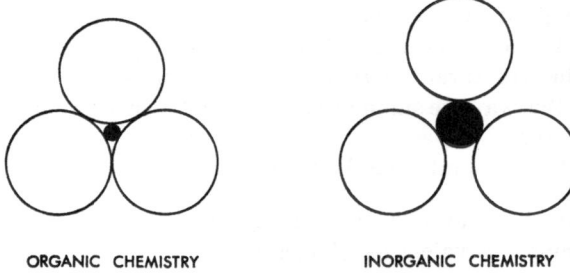

ORGANIC CHEMISTRY   INORGANIC CHEMISTRY

Fig. 37. Shielded and incompletely shielded cores. Models for the discussion of reactivity patterns in small- and large-core chemistry

The major types of bonds encountered in chemistry are summarized schematically in Fig. 38. Broadly speaking, there are two types of atomic cores, small and large, and *three* combinations of core-types: small with small, large with large, and small with large. Significantly, chemists have traditionally recognized *three* types of bonds: covalent, metallic, and ionic. These correspond, respectively, to electrons shared by two or more small cores, two or more large cores, or a small core and a large core.

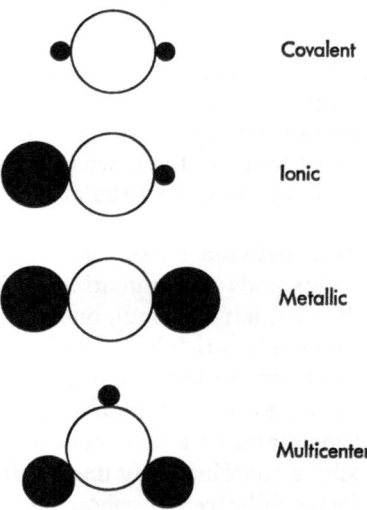

Fig. 38. Bond types. Electron-domain models of chemical bonds

XI. References

[1] *Meyer, V.:* Ber. *27*, 510 (1894).
[2] *Pauling, L.:* Nature of the Chemical Bond, 3rd ed. Ithaca, New York: Cornell Univ. Press 1960.
[3] *Fieser, L. F.:* J. Chem. Educ. *42*, 408 (1965).
[4] *Dreiding, A. S.:* Helv. Chim. Acta *42*, 1339 (1959).
[5] *Brumlik, G. C., E. J. Barrett,* and *R. J. Baumgarten:* J. Chem. Educ. *41*, 221 (1964).
[6] For the use of *rods, disks,* and *hemispheres* to indicate van der Waals interference radii with valence-stick models, see, respectively, ref. [5].
Kooyman, E. C.: J. Chem. Educ. *40*, 204 (1963). — *Larson, G. O.:* J. Chem. Educ. *41*, 219 (1964).
[7] For leading references, particularly to the work of *G. E. Kimball* and co-workers, see *Bent, H. A.:* J. Chem. Educ. *40*, 446 (1963).
[8] *Slater, J. C.:* Phys. Rev. *34*, 1293 (1929).
[9] *Fock, Z.:* Z. Physik *61*, 126 (1930).
[10] *Dirac, P. A. M.:* Proc. Camb. Phil. Soc. *27*, 240 (1931).
[11] *Lennard-Jones, J.:* Proc. Roy. Soc. (London) *198*, 1, 14 (1949).
[12] *Pople, J. A.:* Quart. Rev. (London) *11*, 273 (1957).
[13] *Pauling, L.:* J. Am. Chem. Soc. *53*, 1367 (1931).
[14] *Slater, J. C.:* Phys. Rev. *37*, 481 (1931).
[15] — J. Chem. Phys. *43*, S 11 (1965).
[16] — Phys. Rev. *41*, 255 (1932).
[17] *Hall, G. G.,* and *J. Lennard-Jones:* Proc. Roy. Soc. (London) *202*, 155 (1950).
[18] *Lennard-Jones, J.,* and *J. A. Pople:* Proc. Roy. Soc. (London) *202*, 166 (1950).
[19] *Lewis, G. N.:* J. Am. Chem. Soc. *38*, 762 (1916).
[20] *Lennard-Jones, J.:* J. Chem. Phys. *20*, 1024 (1952).
[21] *Daudel, R., H. Brion,* and *S. Odiot:* J. Chem. Phys. *23*, 2080 (1955).
[22] — Advan. Quantum. Chem. *1*, 115 (1964).
[23] *Edmiston, C.,* and *K. Ruedenberg:* Rev. Mod. Phys. *35*, 457 (1963).
[24] *Foster, J. M.,* and *S. F. Boys:* Rev. Mod. Phys. *32*, 300 (1966).
[25] *Boys, S. F.:* Rev. Mod. Phys. *32*, 296 (1966).
[26] *Adams, W. H.:* J. Chem. Phys. *42*, 4030 (1965).
[27] *Löwdin, P.:* Advan. Chem. Phys. *2*, 207 (1959).
[28] *Clementi, E.:* J. Chem. Phys. *39*, 487 (1963).
[29] *Nesbet, R. K.:* J. Chem. Phys. *43*, S 30 (1965); Advan. Quantum Chem. *3*, 1 (1967).
[30] *Adams, W. H.:* J. Chem. Phys. *34*, 89 (1961).
[31] *Ruedenberg, K.:* Mod. Quantum Chem. *1*, 85 (1965).
[32] *Edmiston, K.,* and *K. Ruedenberg:* In: Quantum Theory of Atoms, Molecules, Solid State, p. 263; *Löwdin, P.,* ed. New York: Academic Press, Inc. 1966.
[33] *Peters, D.:* J. Chem. Soc. 2003, 2015, 4017 (1963); 2901, 2908, 2916 (1964); 3023 (1965); 644, 652, 656 (1966); J. Chem. Phys. *46*, 4427 (1967).
[34] *Pitzer, R. M.:* J. Chem. Phys. *41*, 2216 (1964).
[35] *Klessinger, M.:* J. Chem. Phys. *43*, S 117 (1965).
[36] *Kaldor, U.:* J. Chem. Phys. *46*, 1981 (1967).
[37] *Magnasco, V.,* and *A. Perico:* J. Chem. Phys. *47*, 971 (1968); *48*, 800 (1968).
[38] *Bonaccorsi, R., C. Petrongolo, E. Scrocco,* and *J. Tomasi:* J. Chem. Phys. *48*, 1500 (1968).
[39] *Letcher, J. H.,* and *T. H. Dunning:* J. Chem. Phys. *48*, 4538 (1968).
[40] *Trindle, C.,* and *Sinanoglu:* J. Chem. Phys. *49*, 65 (1968).

41) *Hoyland, J. R.:* J. Am. Chem. Soc. *90*, 2227 (1968).
42) *Frost, A. A.:* J. Chem. Phys. *47*, 3707, 3714 (1967); J. Phys. Chem. *72*, 1289 (1968).
43) —, *B. H. Prentice III*, and *R. A. Rouse:* J. Am. Chem. Soc. *89*, 3064 (1967). — *Frost, A. A.*, and *R. A. Rouse:* J. Am. Chem. Soc. *90*, 1965 (1968).
44) *Edmiston, C.*, and *K. Ruedenberg:* J. Chem. Phys. *43*, S 97 (1965).
45) *Thompson, H. B.:* Inorg. Chem. *7*, 604 (1968).
46) *Gamba, A.:* Phys. Letters *24A*, 64 (1967).
47) *Bent, H. A.:* J. Chem. Educ. *45*, 768 (1968).
48) *Warren, B. E.:* Z. Krist. *72*, 493 (1929).
49) *Bent, H. A.:* J. Chem. Educ. *42*, 348 (1965).
50) — J. Chem. Educ. *43*, 170 (1966).
51) *Wells, A. F.:* Structural Inorganic Chemistry, 3rd ed. London: Oxford Univ. Press 1962.
52) *Ihde, A. J.:* The Development of Modern Chemistry. New York: Harper and Row 1964.
53) *Bent, H. A.:* J. Chem. Educ. *40*, 523 (1963).
54) a) *Lewis, G. N.:* Valence and the Structure of Atoms and Molecules. Chem. Catalogue Co., Inc. 1923, and New York: Dover reprint, Dover Publications, Inc. 1966;
   b) Trans. Faraday Soc. *19*, 454 (1923);
   c) J. Franklin Inst. *226*, 295 (1938).
55) *Sidgwick, N. V.*, and *H. M. Powell:* Proc. Roy. Soc. (London) *176*, 153 (1940).
56) *Gillespie, R. J.*, and *R. S. Nyholm:* Quart. Rev. (London) *11*, 339 (1957).
57) a) *Gillespie, R. J.:* Can. J. Chem. *38*, 818 (1960);
   b) J. Chem. Educ. *40*, 295 (1963);
   c) Angew. Chem. Intern. Ed. Engl. *6*, 819 (1967).
58) *Bragg, W. L.*, and *J. West:* Proc. Roy. Soc. (London) *114*, 452 (1927).
59) *Bernhard, R.:* Sci. Res. *20*, 25 (1969).
60) *Langmuir, I.:* J. Am. Chem. Soc. *41*, 868 (1919).
61) — Chem. Metallurgical Eng. *24*, 533 (1921).
62) *Pauling, L.*, and *S. B. Hendricks:* J. Am. Chem. Soc. *48*, 641 (1926).
63) *Latimer, W.:* J. Am. Chem. Soc. *51*, 3185 (1929).
64) *Bent, H. A.:* Chem. Rev. *61*, 275 (1961).
65) — J. Chem. Educ. *37*, 616 (1960).
66) Cf. *Trindle, C.*, and *O. Sinanoglu:* J. Am. Chem. Soc. *91*, 853 (1969), esp. p. 857.
67) The same descriptive phrases have been used to describe "the moves of silicate chess" by *Lacy, E. D.:* Acta Cryst. *18*, 141 (1965).
68) *Nishikawa, T., T. Itoh*, and *K. Shimoda:* J. Chem. Phys. *23*, 1735 (1955). — *Lide, D. R., Jr.:* J. Chem. Phys. *27*, 343 (1957).
69) *Venkateswarlu, P.*, and *W. Gordy:* J. Chem. Phys. *23*, 1200 (1955).
70) *Blukis, U., P. H. Kasai*, and *R. J. Meyers:* J. Chem. Phys. *38*, 2753 (1963).
71) *Pierce, L.*, and *M. Hayashi:* J. Chem. Phys. *35*, 479 (1961).
72) *Goldschmidt, V. M.:* Trans. Faraday Soc. *25*, 253 (1929).
73) *Roberts, H. L.:* Quart Rev. (London) *15*, 30 (1961).
74) *Case, J. R.*, and *F. Nyman:* Nature *193*, 473 (1962).
75) *Sidgwick, N. V.:* Ann. Repts. Chem. Soc. (London) *30*, 116 (1933).
76) *Werner, A.:* New Ideas on Inorganic Chemistry, trans. by *Hedley, E. P.* New York: Longmans, Green, and Co. 1911.
77) *Rundle, R. E.:* Surv. Prog. Chem. *1*, 81 (1963).
78) *Bent, H. A.:* J. Chem. Educ. *44*, 512 (1967).

79) *Hassel, O.:* Proc. Chem. Soc. (London) 250 (1957); Mol. Phys. *1*, 241 (1958); The Law of Mass Action, pp. 173—183. Oslo: Det Norske Videnskaps-Akademi I 1964.

80) *Hassel, O.,* and *C. Rømming:* Quart. Rev. (London) *16*, 1 (1962).

81) *Bent, H. A.:* Chem. Rev. *68*, 587 (1968).

82) *Mulliken, R. S.,* and *W. B. Person:* Ann. Rev. Phys. Chem. *13*, 107 (1962).

83) *Hassel, O.,* and *K. O. Strømme:* Acta Chem. Scand. *31*, 275 (1959).

84) *Lippard, S. J.,* and *K. M. Melmed:* J. Am. Chem. Soc. *89*, 3929 (1967).

85) *Guggenberger, L. J.,* and *R. E. Rundle:* J. Am. Chem. Soc. *90*, 5375 (1968). — *Hamilton, W. C.:* Structural Chemistry and Molecular Biology, pp. 467—9; *Rich, A.,* and *N. Davidson.* ed. San Francisco: W. H. Freeman and Co. 1968.

86) *Gillespie, R. J.:* Inorg. Chem. *5*, 1634 (1966).

87) *Bartell, L. S.,* and *K. W. Hansen:* Inorg. Chem. *4*, 1775 (1965).

88) —, and *R. M. Gavin, Jr.:* J. Chem. Phys. *48*, 2460, 2466 (1968).

89) *Foss, O.:* Selected Topics in Structural Chemistry. Oslo: Universitets Forlaget 1967.

90) —, and *S. Husebye:* Acta Chem. Scand. *20*, 132 (1966).

91) *Donaldson, J. D.:* Prog. Inorg. Chem. *8*, 287 (1967).

92) *Magnuson, V. R.,* and *G. D. Stucky:* Inorg. Chem. *8*, 1427 (1969).

93) *Atwood, J. L.,* and *G. D. Stucky:* J. Am. Chem. Soc. *91*, 2538 (1969).

94) *Thomas, E. C.,* and *V. W. Laurie:* J. Chem. Phys. *50*, 3512 (1969).

95) *Baenziger, N. C.,* and *A. D. Nelson:* J. Am. Chem. Soc. *90*, 6602 (1968).

96) *Ryan, R. R.,* and *K. Hedberg:* J. Chem. Phys. *50*, 4986 (1969).

97) *Slater, J. C.:* J. Chem. Phys. *41*, 3199 (1964).

98) *Gillespie, R. J.:* Can. J. Chem. *39*, 318 (1961).

99) *Bent, H. A.:* Chemistry *40*, no. 1, 8 (1967).

100) *McCullough, J. D.,* and *R. E. Marsh:* Acta Cryst. *3*, 41 (1950).

101) *Christofferson, G. D.,* and *J. D. McCullough:* Acta Cryst. *11*, 249, 782 (1958).

102) *Chao, G. Y.,* and *J. D. McCullough:* Acta Cryst. *15*, 887 (1962).

103) *Hope, H.:* Acta Cryst. *20*, 610 (1966).

104) *Chen, M. T.,* and *J. W. George:* J. Am. Chem. Soc. *90*, 4580 (1968).

105) *Lowry, T. M.,* and *G. Jessop:* J. Chem. Soc. 323 (1931).

106) *Gerding, H.,* and *H. Houtgraaf:* Rec. Trav. Chim. *73*, 737 (1954).

107) *George, J. W., N. Katsaros,* and *K. J. Wynne:* Inorg. Chem. *6*, 903 (1967).

108) *Greenwood, N. N., B. P. Straughan,* and *A. E. Wilson:* J. Chem. Soc. A 1479 (1966).

109) *Rolfe, J. A., L. A. Woodward,* and *D. A. Long:* Trans. Faraday Soc. *49*, 1388 (1953).

110) *Wynne, K. J.,* and *J. W. George:* J. Am. Chem. Soc. *87*, 4750 (1965).

111) See comments on CH_3TeBr_3 and CH_3SeCl_3 in ref. [104] and comments on $CsIF_6$ and XeF_6 by *Beaton, S. P., D. W. A. Sharpe, A. J. Perkins, I. Sheft, H. H. Hyman,* and *K. Christe:* Inorg. Chem. *7*, 2174 (1968).

112) For carbon, cf. *Breslow, R., S. Garratt, L. Kaplan,* and *D. LaFolette:* J. Am. Chem. Soc. *90*, 4051, 4056 (1968).

113) *Berry, S.:* J. Chem. Phys. *32*, 933 (1960).

114) *Lindemann, F. A.:* Phil. Mag. *29*, 127 (1915).

115) *Langmuir, I.:* J. Am. Chem. Soc. *38*, 2243 (1916).

116) *Hull, A. W.:* J. Franklin Inst. *193*, 189 (1922).

117) *Thomson, J. J.:* Phil. Mag. *44*, 663 (1922).

118) *Bragg, W. H.:* X-rays and Crystal Structure (4th ed.). London: G. Bell and Sons 1924.

119) *Lennard-Jones, J.:* Proc. Roy. Soc. (London) *120*, 734 (1928).

H. A. Bent

120) *Fajans, K.:* Quanticule Approach to the Metallic State; paper presented at a Gordon Res. Conf. on Metals, New Hampton, N. H., July, 1950.
121) *Houston, W. V.:* Phys. Today *16*, 26 (1963).
122) *Löwdin, P. O.:* Phys. Rev. *97*, 1509 (1955). — *Pratt, G. W., Jr.:* Phys. Rev. *102*, 1303 (1956).
123) *Hylleraas, E. A.:* Z. Physik. *54*, 347 (1929).
124) *Eckart, C.:* Phys. Rev. *36*, 878 (1930).
125) *Dewar, M. J. S.,* and *C. E. Wulfman:* J. Chem. Phys. *29*, 158 (1958). — *Dewar, M. J. S.,* and *N. L. Hojvat:* Proc. Roy. Soc. (London) *264*, 431 (1961).
126) *Linnett, J.:* J. Am. Chem. Soc. *83*, 2643 (1961); The Electronic Structure of Molecules. A New Approach. New York: John Wiley and Sons, Inc. 1964.
127) *Buchler, A., J. L. Stauffer, W. Klemperer,* and *L. Wharton:* J. Chem. Phys. *39*, 2299 (1963).
128) *Gillespie, R. J.:* J. Am. Chem. Soc. *82*, 5978 (1960).
129) *Shull, C. G., W. A. Strauser,* and *E. O. Wollan:* Phys. Rev. *83*, 333 (1951).
130) *Wollan, E. O., H. R. Child, W. C. Koehler,* and *M. K. Wilkinson:* Phys. Rev. *112*, 1132 (1959).
131) *Mathieson, A. M., D. P. Mellor,* and *N. C. Stephenson:* Acta Cryst. *5*, 185 (1952).
132) *Morrow, J. C.:* Acta Cryst. *15*, 851 (1962).
133) *Slater, J. C.:* Phys. Rev. *35*, 509 (1930).
134) *Ramsay, W.:* J. Chem. Soc. *93*, 774 (1908).
135) *Pauling, L.:* Nature of the Chemical Bond, 3rd ed; p. 393. Ithaca, New York: Cornell Univ. Press 1960.
136) *Faraday, M.:* Phil. Mag. *24*, 136 (1844).
137) *Langmuir, I.:* J. Am. Chem. Soc. *41*, 868 (1919).
138) *Rice, O. K.:* J. Chem. Phys. *1*, 649 (1933).
139) *Stokes, R. H.:* J. Am. Chem. Soc. *86*, 2333 (1964).
140) *Raich, J. C.,* and *R. H. Good Jr.:* J. Phys. Chem. Solids *26*, 1061 (1965).
141) *Cutler, M.:* J. Chem. Phys. *46*, 2044 (1967).
142) *Phillips, J. C.,* and *L. Kleinman:* Phys. Rev. *116*, 287 (1959).
143) *Cohen, M. H.,* and *V. Heine:* Phys. Rev. *122*, 1821 (1961).
144) *Austin, B. J., V. Heine,* and *L. J. Sham:* Phys. Rev. *127*, 276 (1962).
145) *Szas, L.,* and *G. McGinn:* J. Chem. Phys. *45*, 2898 (1966).
146) *Austin, B. J.,* and *V. Heine:* J. Chem. Phys. *45*, 928 (1966).
147) *Chang, E. S., R. T. Pu,* and *T. P. Das:* Phys. Rev. *174*, 1 (1968).
148) *Weare, J. H.,* and *R. G. Parr:* Chem. Phys. Letters *1*, 349 (1967).
149) *Hall, G. G.:* Chem. Phys. Letters *1*, 495 (1967).
150) *Rowe, D. J.:* Rev. Mod. Phys. *40*, 153 (1968).
151) *Clementi, E.:* Chem. Rev. *68*, 342 (1968).
152) *Whyte, L. W.:* The Six-Cornered Snowflake; ed. and trans. by *Hardie, C.* Oxford: Oxford Univ. Press 1966.
153) *Slater, J. C.:* Phys. Rev. *34*, 1293 (1929).
154) *Weinreich, G.:* Solids. New York: John Wiley and Sons, Inc. 1965.
155) *Lennard-Jones, J.:* Advan. Sci. *11*, 136 (1954).
156) *Weisskopf, V. F.:* Science *149*, 1181 (1965).
157) *Petersen, A.:* Quantum Mechanics and the Philosophical Tradition. Cambridge, Mass.: M. I. T. Press 1968.

Received September 9, 1969

Non-Destructive Techniques in Activation Analysis

Prof. W. D. Ehmann

University of Kentucky, Department of Chemistry, Lexington, Kentucky and
Arizona State University, Department of Chemistry, Tempe, Arizona, USA

Contents

I.   Introduction 49

II. Some Non-Destructive Activation Methods 52

    A. 14 MeV Neutron Activation Analysis 52
       1. General 52
       2. Neutron Generators 55
       3. Sample Handling and Packaging 58
       4. Precision and Accuracy 59
       5. Some Applications 62

    B. Use of Ge(Li) Detectors for Multi-Element Trace Analysis 65
       1. General 65
       2. Some Applications 67

    C. Applications of Gamma-Gamma Coincidence Spectrometry 71
       1. General 71
       2. Some Applications 79

    D. Other Non-Destructive Techniques 81
       1. General 81
       2. Photoactivation Analysis 81
       3. Charged Particle Activation Analysis 82
       4. Prompt Gamma-Ray Analysis 83
       5. Neutron Activation Followed by Delayed Neutron Counting 84

III. Summary 84

IV. References 85

I. Introduction

The technique of activation analysis evolved from the work of *Hevesy* and *Levi* in 1936 on the reactions of neutrons with rare earth elements [1]. The technique was not widely used until 1950, when a period of rapid growth in the number of applications started. The literature in the field today approaches 4000 publications and monographs, including nearly 600 contributions in 1967 alone. An excellent bibliography cross-indexed by author, element determined, matrix analyzed and technique used has been recently published by the U.S. National Bureau of Standards [2].

Numerous books and monographs published in the last decade are available to provide fundamental information on the method [3-17].

The technique permits qualitative and quantitative elemental analyses in a wide variety of sample matrices. Principal advantages of the method are

> high sensitivity (often in the sub-nanogram region) for many elements,
> freedom from laboratory and reagent contamination problems,
> and in many cases the possibility of rapid and essentially non-destructive analyses.

The method yields only gross sample elemental abundances and by itself does not permit distinguishing between different chemical states of an element in mixtures. For many elements, activation analysis is the most sensitive analytical technique available. It is the purpose of this paper to show that recent advances in instrumentation now make possible accurate, rapid, and often non-destructive determinations of elemental abundances by this technique and that it should be included among the routine analytical tools available in any modern analytical facility.

Activation analysis is based on the production of radioactive nuclides by means of induced nuclear reactions on naturally occurring isotopes of the element to be determined in the sample. Although irradiations with charged particles and photons have been used in special cases, irradiation with reactor thermal neutrons or 14 MeV neutrons produced by Cockcroft-Walton type accelerators are most commonly used because of their availability and their high probability of nuclear reaction (cross section). The fundamental equation of activation analysis is given below:

$$A = n\theta\sigma(1-e^{-\lambda t_i})e^{-\lambda t_d} \tag{1}$$

where,

A = the activity of the product radionuclide (disintegrations per second).
n = the number of atoms of the target nuclide in the sample.
θ = "flux", or the area-time density of the bombarding particles (particles per square centimeter per second).
σ = "cross section", or the probability that a target nucleus will undergo a specific nuclear reaction with the incident particles (square centimeters).
λ = the decay constant of the product radionuclide (ln $2/H$, where H is the half-life of the radionuclide in time units identical to those used for t_i and t_d).
t_i = the time of irradiation of the sample.
t_d = the time of decay following irradiation and prior to the start of counting.

The term $(1-e^{-\lambda t_i})$ is referred to as the saturation factor and approaches unity as the time of irradiation becomes large with respect to the half-life of the product radionuclide. Obviously, little is gained by irradiation of the sample for periods longer than 2 or 3 times the half-life of the product radionuclide.

The weight (W) of the element in the sample in units of grams may then be computed from Eq. 2:

$$W = n \, (\text{At.Wt.})/6.02 \times 10^{23} \, (\text{Iso.Abd.}) \tag{2}$$

where,

At. Wt. = the atomic weight of the element
Iso. Abd. = the isotopic abundance of the target nuclide expressed as a fraction.

The use of Eq. 1 depends on accurate information on the flux, the cross section, and the absolute activity of the product radionuclide. In practice these data are difficult to obtain and, in addition, the flux may not be constant with respect to time or position of irradiation.

It is the usual practice, therefore, to irradiate a standard containing a known amount of the element to be determined along with the samples and to count both standards and samples with the same detector system. In this case the absolute value of the flux, the constancy of the flux, and the detection efficiency of the detector do not enter into the determination. By combining the standard activation equations for both standard and sample, the following equation results:

$$\frac{R_{\text{std}}}{R_{\text{sam}}} = \frac{W_{\text{std}} \, (e^{-\lambda t_d})_{\text{std}}}{W_{\text{sam}} \, (e^{-\lambda t_d})_{\text{sam}}} \tag{3}$$

where,

R_{std} and R_{sam} = the experimental counting rates for the standard and sample, respectively.
W_{std} and W_{sam} = weights of the element in the standard and sample, respectively, assuming identical isotopic composition.

Since the standard and sample are to be counted with the same detector system, they are ordinarily counted in sequence and the decay factors ($e^{-\lambda t_d}$) will not be identical.

When the indicator radionuclide has a half-life short with respect to the counting time, as in the determination of oxygen via 14 MeV neutron activation to produce 7.37 sec half-life [16]N, additional corrections to the above equations are required [18].

Many of the early applications of activation analysis were designed specifically to take advantage of its high sensitivity for rarer elements and its freedom from sources of contamination. In these trace element studies time-consuming radiochemical separations employing inactive carriers were often required to obtain the product radionuclide in a radiochemically pure form for counting. This was necessary in activation analysis of complex matrices, due to the production of macro-amounts of interfering activities from major abundance components of the sample. Non-destructive determination of several major elements in complex matrices was sometimes possible by means of direct gamma-ray scintillation spectrometry using a NaI(Tl) detector [19]. Unfortunately, the energy resolution of most commercially available NaI(Tl) crystals is only about 50 KeV at 0.7 MeV (full width at half maximum for a gamma-ray photopeak). Hence, the use of NaI(Tl) detectors often required use of computer methods of spectrum resolution to resolve the overlapping photopeaks.

Recently, the development of extremely high resolution solid state Ge(Li) gamma-ray detectors, the application of gamma-gamma coincidence techniques, and the availability of low cost 14 MeV neutron generators have renewed interest in activation analysis as a means for routine rapid and non-destructive elemental abundance determinations of both major and trace elements. Some of these new techniques will be discussed in the sections that follow.

II. Some Non-Destructive Activation Methods

A. 14 MeV Neutron Activation Analysis

1. General

In many chemistry laboratories today it is possible to find small low energy particle accelerators especially designed to function as generators of 14 MeV neutrons for analytical work. Much of the early emphasis in 14 MeV neutron activation was directed towards the *determination of oxygen* in pure metals, geological materials and meteorites [20–23]. This was natural, since direct oxygen determinations have long been a difficult problem for the analytical chemist. Oxygen determinations with 14 MeV neutrons via the $^{16}O(n,p)^{16}N$ reaction have been shown to be rapid, accurate, and suitable for large scale semi-automated analytical applications [24,26].

Many analysts are not aware that over *one-half of the elements in the periodic table may be determined* at levels of one milligram, or less, with commercially available neutron generators. Actual experimental sensitivities are, of course, dependent on levels of interfering activities generated in the activation of complex matrices. The sensitivities listed in

Fig. 1 are based on a routine operating flux of $2 \cdot 10^8$ neutrons cm^{-2} sec^{-1}. With a fresh tritium target a useful 14 MeV neutron flux of $2.5 \cdot 10^9$ neutrons cm^{-2} sec^{-1} may be attained for a short period of time. However, depletion of the target is rapid when operating with the high beam currents required. For longer-lived indicator radionuclides irradiation times longer than 5 minutes could be used to improve sensitivities. Again, target depletion usually restricts the routine use of 14 MeV activation to relatively short irradiation periods. For special applications, most of the sensitivity limits listed in Fig. 1 could easily the improved by a factor of ten.

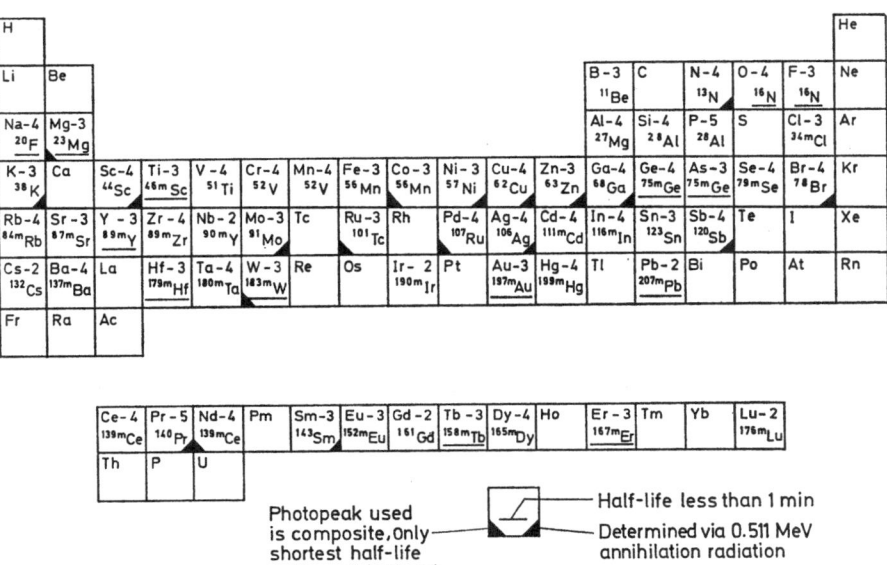

Fig. 1. Sensitivities for 14 MeV neutron activation analysis based largely on the compilation of *Cuypers* and *Cuypers* [25]. The symbol Mn —4 signifies a sensitivity of 10^{-4} grams for Mn using a 14 MeV neutron flux of approximately 2×10^8 neutrons-cm^{-2} sec^{-1}. The limit of sensitivity was considered to be the point at which the integrated photopeak counts equally three standard deviations of the background count in the photopeak region, using the baseline method. Only the indicator radionuclide resulting in the highest sensitivity is listed. In many cases alternate indicator radionuclides could be used. For indicator radionuclides with half-lives greater than one minute the irradiation, delay and counting times were 5 minutes, 1 minute and 5 minutes, respectively. For shorter half-life indicator radionuclides these times were 3 half-lives, 1 half-life and 3 half-lives, respectively. Detectors were two 3″ × 3″ NaI(Tl) crystals 2.4 cm apart with the sample centered between them

In 14 MeV neutron activation analysis the four principal nuclear reactions leading to the formation of indicator radionuclides are as follows:

Target Nuclide	Nuclear Reaction	Product Radionuclide
^AZ	(n,p)	$^A(Z-1)$
^AZ	(n,α)	$^{A-3}(Z-2)$
^AZ	$(n,2n)$	^{A-1}Z
^AZ	(n,n')	$^{A*}Z$

The *cross sections* for (n,γ) reactions common in reactor thermal neutron activation generally decrease with increasing neutron energy with the exception of resonance-capture cross section peaks at specific energies. This reaction is, therefore, not important in most 14 MeV activation determinations. However, some thermalization of the 14 MeV flux may always be expected due to the presence of low Z elements in the construction materials of the pneumatic tubes, sample supports, sample vial, or the sample itself (particularly when the sample is present in aqueous solution). The elements Al, Mn, V, Sn, Dy, In, Gd, and Co, in particular, have high thermal neutron capture cross sections and thermal capture products have been observed in the 14 MeV neutron irradiation of these elements in spite of care taken to reduce the amount of low Z moderating materials in the region of the sample irradiation position [25].

The $(n,2n)$, (n,n') and most (n,p) reactions on stable nuclides are endoergic; that is, they have a non-zero reaction energy threshold. The (n,α) reaction is endoergic for low Z target nuclides, but becomes exoergic for higher Z nuclides. In most cases, cross sections for the four principal 14 MeV neutron reactions are less than 1 barn (10^{-24} cm^2). In contrast, the thermal neutron (n,γ) reaction is always exoergic and may have cross sections exceeding 10^5 barns.

It is obvious, therefore, that 14 MeV neutron activation analysis can not compete with thermal neutron activation analysis as a technique for trace element analysis. In simple matrices, however, the rapid and non-destructive nature of the technique recommends its use for routine analysis of large numbers of samples for elemental abundances at the one milligram level, or above. It is unfortunate that the element carbon can not be determined by this technique. The nuclear reaction $^{12}C(n, 2n)^{11}C$ which would be of great analytical importance is endoergic to the extent of nearly 19 MeV. This reaction is obviously not energetically possible using the 14.7 MeV neutrons produced by the $^2H(^3H,n)^4He$ reaction commonly employed in most neutron generators.

It should be noted that commercial neutron generators are also easily adopted to the generation of 2.8 MeV neutrons produced by the $^2H(^2H,n)^3He$ reaction. In most cases it is merely necessary to replace the tritium target with one containing occluded deuterium. The neutron yield from this reaction is much less than for the D—T reaction and the useful flux is often not much greater than could be obtained by use of isotopic sources. About 35 elements have been found to possess reasonably high $(n,n'\gamma)$ or (n,γ) cross sections for 2.8 MeV neutrons [41]. Since the 8 most common elements in the earth's crust are not among those readily activated, there is some potential application of 2.8 MeV neutrons in analyses for certain elements in minerals and ores, where major element interferences via 14 MeV activation may be a problem.

2. Neutron Generators

Commercial 14 MeV neutron generators available today are largely simple particle accelerators of the Cockcroft-Walton type [27]. Van de Graaff type accelerators have also been used for activation analysis, but are generally regarded as too expensive to be used exclusively for this purpose. The Cockcroft-Walton type of neutron generator consists essentially of an ion source capable of producing atomic or molecular deuterium ions (D^+ or D_2^+), acceleration electrodes supplied by voltage-multiplier or transformer-rectifier circuits, a drift tube maintained at a high vacuum, and a tritium-containing target. A schematic diagram of a simple Cockcroft-Walton neutron generator is shown in Fig. 2.

Three types of ion sources are in common use. The *radio frequency ion source* [28] has advantages of a high (70—90%) atomic deuterium ion beam, high stability, and low deuterium gas consumption. This source is, however, very sensitive to impurities in the deuterium gas and requires

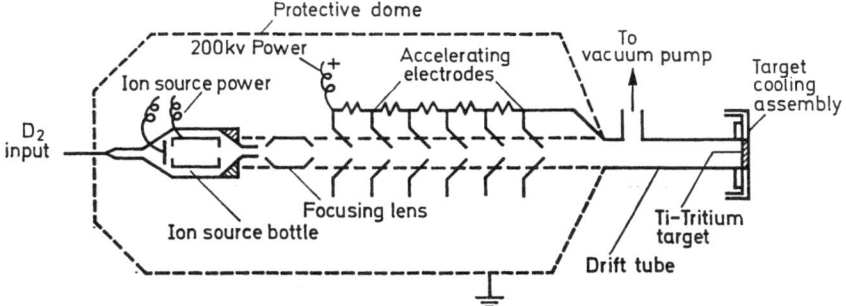

Fig. 2. Schematic diagram of the major components of a Cockcroft-Walton type neutron generator

more attention to obtain optimum performance. The deuterium is usually introduced into this source by diffusion through a heated palladium metal cylinder. The rate of introduction of deuterium into the source may be easily controlled by varying the heating temperature of the palladium cylinder. The palladium leak is to be preferred over a simple mechanical type of gas input regulation, since it contributes to a cleaner vacuum system by preventing introduction of impurities into the ion source. Impurity deposits on the surface of the target which would reduce the neutron yield are minimized by maintenance of a clean vacuum system.

The *Penning ion source* achieves ionization of the deuterium gas by means of acceleration of streams of electrons released by cathodes on either side of a hollow anode chamber. Its advantages of simplicity of construction and operation, long operating lifetime, and ability to generate moderately high beam currents are balanced by its low efficiency in producing atomic D^+ ions. Ordinarily, over 80% of the beam produced by the Penning ion source is made up of molecular D_2^+ ions. An acceleration potential of 200 kilovolts will accelerate both a D^+ ion and a D_2^+ ion to an energy of 200 KeV, but each deuteron in the molecular ion acquires only $1/2$ the energy, or 100 KeV. The 14 MeV neutron yield of the generator in units of neutrons sec^{-1} milliampere^{-1} is a function of acceleration voltage, as shown in Fig. 3. Since the neutron yield of a 100 KeV deuteron is only approximately 0.36 that of a 200 KeV deuteron,

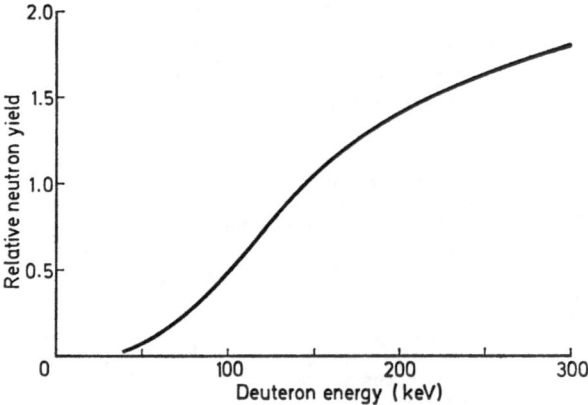

Fig. 3. Variation of the relative 14 MeV neutron yield at a constant beam current as a function of deuteron energy. It should be noted that acceleration of only molecular D_2^+ ions through a potential of 200 KV results in each deuteron in the molecular ion having a energy of 200/2, or only 100 KeV. This calculated curve is based on use of a thick Ti-T target [100]

the overall 14 MeV neutron yield of a pure molecular ion beam (D_2^+) accelerated through a potential of 200 KV will be only 0.72 of that which could be obtained by a pure atomic ion beam (D^+) accelerated through the same potential.

The cross section for the $^3H(d, n)^4He$ reaction reaches its maximum value at only 107 KeV incident deuteron energy. When "thick" ($\sim$ 1 mg cm^{-2} thick deposit of titanium) titanium-tritium targets are used, however, the neutron yield continues to increase even above 200 KV acceleration potential. This is due to increased penetration of the deuteron beam into the tritium enriched layer. Since the penetration of molecular deuterium ions is less than that for monatomic deuterium ions for the same acceleration potential, accelerators using Penning ion sources require extremely clean vacuum systems to minimize build-up of deuteron absorbing deposits on the surface of the target.

More recently the modified *duoplasmatron type of ion source* [29] has been used with neutron generators. In this source both electrostatic and magnetic focusing are employed to generate the plasma in high density in the restricted region of the beam extraction port. These sources are capable of generating extremely high beam currents with high power efficiency and low gas consumption. Atomic deuterium ion yields of at least 60% are possible with this source. The principal disadvantage of this source at present is its high cost.

The targets used in neutron generators consist of tritium occluded in a thin layer of titanium deposited on a copper backing. Since the deuteron beam generates an appreciable amount of heat in the target, the back copper face of the target is usually cooled by circulating water or Freon-113. The neutron yield from the $^3H(d, n)^4He$ reaction is very nearly isotropic and the product neutron energy varies from approximately 14.7 MeV in the forward direction to approximately 13.4 MeV at 180° from the direction of the incident deuteron beam when using an acceleration potential of 150 KV. Tritium is evaporated from the target during deuteron bombardment and only a small fraction of the total amount of tritium in the target is effectively used in neutron generation. In the operation of a neutron generator a rather rapid decrease in neutron yield is observed over the first few minutes of operation, followed by an approximately exponential decrease in yield with a half-time of approximately 3—4 milliampere-hours of target exposure. In our laboratories targets are changed about every 15 milliampere-hours.

The necessity of frequent target changes and the attendant risk of tritium contamination of personnel and laboratories [30] has led to the development of high yield *sealed-tube generator systems* in which the tritium supply is constantly replenished [31-33]. Ordinarily these systems operate by accelerating a mixture of deuterium and tritium ions into the target and recirculation of the gases through a Penning type ion source. Neutron yields in excess of 10^{11} neutrons sec^{-1} with little or no degradation in the yield over periods of operation in excess of 100 hours have been obtained. In addition to the safety and convenience factors, the sealed-tube generator systems are generally much smaller than the conventional pumped-type accelerators which permits greater flexibility in shielding arrangements. It has been suggested that sealed-off neutron tubes with outputs above 10^{12} neutrons sec^{-1} and neutron yield half-times of several hundred hours are feasible [33]. These generators are particularly attractive for installation in industrial facilities where analyses are performed on a routine basis by technician-level personnel. More detailed reviews of accelerator systems for activation analysis are available in the literature [36,37] and are also available from manufacturers of neutron generators.

3. Sample Handling and Packaging

The majority of the applications of 14 MeV neutron activation analysis involve the use of short-lived indicator radionuclides. Therefore, it is essential that the sample be returned quickly to the counting station following irradiation. Pneumatic sample transfer systems employing compressed nitrogen, or a vacuum are most commonly used [34,35]. An inexpensive system may be constructed from ordinary low density polyethylene tubing [18]. Irradiation, delay and counting times are ordinarily controlled by means of preset timing circuits. Completely automated control and transfer systems are available commercially.

Samples for analysis are ordinarily packaged in machined or heat-sealed [38] polyethylene vials ("rabbits") for 14 MeV neutron irradiation. Sample sizes of 0.1 to 2 grams are most often used, although samples weighing in excess of 25 grams have been used in special cases. The small sample size is desirable in order to minimize the effect of neutron flux inhomogeneity caused by sample self-absorption and flux variations across the sample irradiation position. For the highest precision, biaxial rotation of the sample both in the irradiation and the counting positions is desirable. *Dual biaxial rotation assemblies* are commercially available. These systems provide for simultaneous irradiation of the standard and the sample and, hence, improve precision in the analyses by minimizing errors due to time variations in the intensity of the neutron flux [39]. Severe sample inhomogeneity, or situations where sample and standard vary appreciably in physical size or composition may lead to a lack of precision even when a biaxial rotation assembly is used [40].

For the analysis of rocks and minerals, the sample is usually powdered and the standard is prepared by doping *Spec-Pure* SiO_2 with an aliqout of a standard solution of the element to be determined. For the analysis of solutions, the standard is usually an aliquot of a standard solution of the element to be determined diluted to a volume approximating that of the sample to be analyzed. Metal samples are commonly irradiated in the form of filings, or small preformed cylinders. In this case, standards of independently analyzed metal would be desired. In some cases it might be more desirable to dissolve the metal, so that easily prepared solution standards could be used. As stated previously, efforts to achieve physical and gross chemical similarity of standard and sample are required for analyses of the highest quality.

It is common to prepare standards that contain approximately the same or slightly larger amounts of the element of interest than the amount estimated to be found in the sample. In this way errors associated with varying sample and standard instrumental dead-time corrections in the counting system are minimized. This is an important consideration when counting indicator radionuclides having half-lives short with respect to the time of counting.

Although commercially available polyethylene tubing and vials that are used for the construction of "rabbits" are relatively free of metal contamination, *blanks on empty irradiation vials* should always be run. Small amounts of oxygen absorbed or included in ordinary polyethylene can offer a serious interference in the determination of oxygen in small samples. "Oxygen-free" polyethylene prepared in a nitrogen atmosphere can be obtained commercially. The dies used to form some polyethylene items apparently result in slight contamination of the surfaces of these items with chromium. Pre-irradiation cleaning with acids has been shown to be desirable for materials used in the construction of the "rabbits".

4. Precision and Accuracy

Discussions of errors associated with the technique of activation analysis in general may be found in many of the books and monographs referenced in the introduction to this paper. Interferences unique to 14 MeV neutron activation techniques have been reviewed by *Mathur* and *Oldham* [42] and a discussion of precision has been published by *Mott* and *Orange* [43].

Some of the most important factors affecting the precision and the accuracy of 14 MeV activation are reviewed below. The random errors as listed in the section on precision are generally reduced in importance by running replicate analyses on each of several aliquants of the sample and averaging all the results. Consideration given to the reduction of these sources of error will, of course, result in a reduction of the number of replicate determinations required to obtain results of the desired accuracy.

Some Factors Affecting Precision

(1) *Non-reproducible positioning* of sample and/or standard during irradiation and counting. Use of a biaxial dual capsule rotator and rotation of the capsules during counting are commonly used procedures in minimizing this source of error.

(2) *Non-uniform flux distribution* across the sample irradiation position. This is an important consideration in the analysis of non-homogeneous samples, or samples and standards that differ appreciably in their physical dimensions and properties. An isotropic flux distribution from the D—T reaction is only realized at a distance of several inches from the tritium target [40]. The error may be minimized by increasing the distance from the sample irradiation position to the generator target. This will, of course, reduce the sensitivity of the determination due to the reduced flux level at the greater distance. In general non-homogeneous samples should be avoided and standards physically similar to the samples should be used. Use of a defocused deuteron beam is also helpful [43].

(3) *Time variations in the intensity* of the flux during irradiation. This is an important consideration only when a single sample transfer system is used. Gas-filled BF_3 neutron counter tubes are often used to monitor the neutron flux in order to normalize the data when the sample and the standard are not irradiated simultaneously. Gain shifts and dead-time effects associated with the use of neutron monitoring detectors also contribute to the errors associated with a single sample transfer system.

(4) *Errors in photopeak baseline selection* and photopeak integration. While manual methods depending on subjective judgement are commonly used for baseline selection (the baseline is often assumed to be approximately linear over a small number of channels), analyses of multicomponent spectra often require mathematical curve-fitting and resolution with the aid of a computer.

(5) *Errors in timing.* For the determination of elements yielding short half-life indicator radionuclides (such as in the determination of oxygen via 7.3 sec ^{16}N), accurate timing is extremely important. For these cases electronic scaler timers are to be preferred over electromechanical types of timers. Errors due to variable detector dead-time must also be considered when the gross activities of the sample and the standard differ appreciably and the indicator radionuclide is short-lived.

(6) *Errors in standard and sample preparation.* Gravimetric and volumetric errors in the preparation of the standard and the sample are

common to all methods of analysis. When using the small samples common in routine activation analysis, care must be taken to assure that the sampling procedure used is valid. Homogenization of the bulk sample prior to sampling and use of replicate samples is desirable.

(7) *Counting statistics.* The statistical standard deviation of the number of counts in the photopeak is equal to the square root of the integrated photopeak counts. For most useful determinations it is assumed that sufficient activity is produced in order to assure errors due to counting statistics are less than 1%.

(8) *Instrumental errors in counting.* Since these errors may occur to either sample or standard, the overall effect on the determination can in some sense be regarded as random in nature. Pulse pile up at high counting rates, analyzer channel drop, and gain shift where integration is done between preset limits may all contribute to an erroneously low observed counting rate.

Some Factors Affecting Accuracy

(1) *Primary interference reactions.* These are nuclear reactions on elements other than the element to be determined which yield the same indicator radionuclide. For example, silicon is determined by the $^{28}Si(n,p)^{28}Al$ reaction. However, the same indicator radionuclide is produced from phosphorus by the $^{31}P(n,\alpha)^{28}Al$ reaction. Hence, a high phosphorus abundance in a sample will lead to erroneously high values for silicon. Corrections may be applied to the data if concentrations of the interfering elements can be determined independently.

(2) *Secondary interference reactions.* These are nuclear reactions induced by secondary particles produced in the sample or its immediate environment which will produce the indicator radionuclide by interaction with elements other than the one to be determined. For example, nitrogen is usually determined by the $^{14}N(n,2n)^{13}N$ reaction. The 14 MeV neutrons may generate recoil protons by collision with hydrogen atoms in the vial, transfer tubes, or sample support assembly. These recoil protons may induce the $^{13}C(p,n)^{13}N$ reaction with the carbon of the vial, leading to the formation of the same indicator radionuclide. This type of interference is ordinarily not serious in cases other than the nitrogen determination.

(3) *Gamma-ray spectral interferences.* This is important in cases where the 0.511 MeV annihilation radiation from the product radionuclide is measured in the determination. Obviously, all other reactions yielding positron emitters would provide an interference. In the determination of oxygen a spectral interference is produced if the sample contains an appreciable amount of boron. The ^{11}Be produced by the $^{11}B(n,p)^{11}Be$ decays with the emission of an 11 MeV negatron and also gamma-rays

having energies close in energy to those emitted by the [16]N indicator radionuclide used in the determination of the oxygen.

(4) *Self-shadowing and resonance capture effects.* The use of small samples and standards so that the neutron flux is not appreciably attenuated between the exterior and interior of the irradiation unit is to be desired. When large samples are used or appreciable high cross section material is present in the matrix, it is important that the standard be prepared with a matrix physically and chemically similar to that of the sample.

(5) *Inaccurate calibration of relative counting efficiencies.* In the use of a dual sample transfer system where both sample and standard are counted simultaneously with different counters it is important that the relative efficiencies of the two counting systems be accurately determined and that these efficiencies do not vary with time. These errors may be partly compensated for by irradiating the same sample and standard several times and alternately reversing the sample and standard counting positions.

A reported relative standard deviation of 1—2% for small sets of replicate analyses is common in many recent papers employing this technique. Results falling within 0.10% of the theoretical percentage of oxygen in samples containing about 50% oxygen based on 32 replicate determinations have been reported [24]. For the 32 determinations the relative standard deviation for the data was approximately 0.3%. This level of accuracy and precision is certainly adequate for most analytical purposes.

5. Some Applications

As mentioned previously, one of the early applications of 14 MeV neutron activation analysis was in the analysis of rocks and stony meteorites for oxygen and silicon. Both of these elements are rather difficult to determine accurately by conventional analytical techniques. Fig. 4 illustrates the gamma-ray spectrum obtained at three different decay times following 14 MeV neutron irradiation of a stony meteorite containing approximately 35% oxygen and 18% silicon. The sample size was approximately 200 mg. Oxygen was determined by integration of the counts above 4.5 MeV. The activity in this energy region is almost entirely due to the photopeak of [16]N and its two escape peaks hence, no baseline determination or background determination is required [23]. After a decay period of approximately 1 minute silicon was determined by integrating the 1.78 MeV photopeak of [28]Al produced from silicon [44]. A standard rock sample with well established contents of silicon and oxygen may be used to provide simultaneously a standard for both elements. Since irradiation,

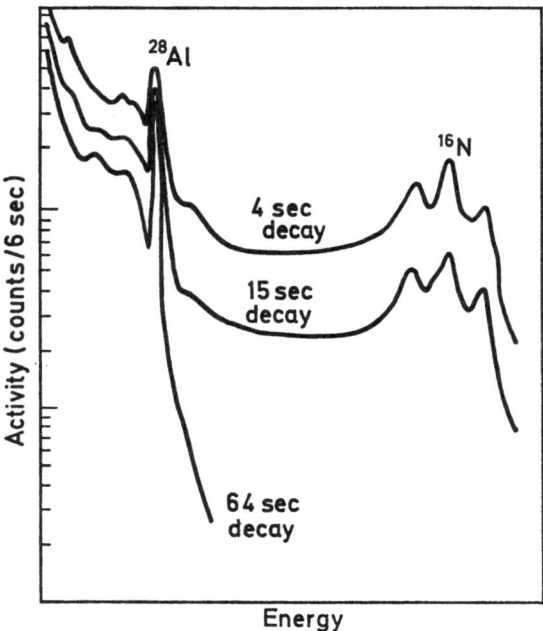

Fig. 4. Gamma-ray spectrum of an irradiated stony meteorite following a 13 sec irradiation with a flux of approximately 10^8 14 MeV neutrons $cm^{-2} sec^{-1}$. Gamma-ray photopeaks of ^{28}Al and ^{16}N produced by activation of silicon and oxygen, respectively, are prominent features of the spectrum. The features at 5.62 and 5.11 MeV are the first and second escape peaks due to pair production events of the primary 6.13 MeV ^{16}N gamma-rays in the $3'' \times 3''$ NaI(Tl) detector

delay and counting times total less than 5 minutes ,a large number of samples may be processed in a working day, as compared to methods based on conventional techniques.

An interesting modification of this *technique for oxygen* has been developed by *Morgan* [49] in our laboratories. A 4096-channel analyzer is used in its multiscaler mode to sequentially monitor the relative neutron flux (via a BF_3 neutron counter), the decay of the activated sample, and the irradiation, delay and counting times. The analyzer is activated at the start of the "rabbit" send cycle and at the start of the irradiation the output of the BF_3 neutron monitor is fed into the analyzer memory. The relative neutron flux is measured from this record and any time variations in the flux during irradiation are easily detected and may be partly compensated for by use of an appropriate computer program. Since the multiscaler timing sequence is continuous, irradiation, delay, and counting times may be read directly from the analyzer output. During the delay period in which the sample is transferred to the counting station, the multiscaler input is automatically switched to the output of a $4'' \times 4''$ NaI(Tl) well crystal. At the start of the counting cycle the activity of the sample is measured by the analyzer as a function of time. A least squares fit may be applied to the decay curve by the computer and the net ^{16}N counts for the counting period calculated. A typical analyzer output trace is shown in Fig. 5. The energy region of interest is selected by a single channel analyzer inserted between the detector and the multichannel analyzer. This technique has the advantage of recording all data necessary for the computations on a single computer compatible output tape. As stated previously, variations in the neutron flux and the presence of longer and shorter half-life impurity activities may be corrected for by application of appropriate computer programs.

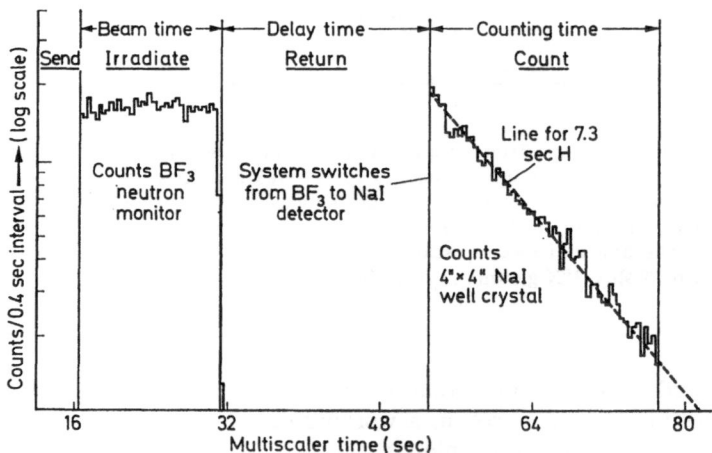

Fig. 5. Use of a 4096-channel analyzer in the multiscaler mode for the determination of oxygen via 14 MeV neutron activation analysis

Activation with 14 MeV neutrons has been used to determine the oxygen content of various metals such as beryllium [20], Cl, F, O, Na, Si, and various rare earths in complex molten salt electrolytes [45], the protein content of food products by means of the nitrogen content [46], $^{16}O/^{18}O$ and $^{14}N/^{15}N$ isotopic ratios in stable isotope tracer experiments [47,48], and in a wide variety of other applications. One application we have found to be of use in our own laboratory is in the rapid determination of the carrier chemical yields in trace element determinations via reactor irradiation and radiochemical separations. Reactivation of the purified carrier-containing sample with 14 MeV neutrons in most cases will produce different short-lived indicator radionuclides than are produced in the original thermal neutron irradiation. Hence, with the use of the appropriate standards it is easy to determine the chemical yield of the separation procedure without converting the separated element to an acceptable gravimetric weighing form. In conventional separation procedures the conversion of the element to the final weighing form is often the most time consuming step in the analysis.

B. Use of Ge(Li) Detectors for Multi-Element Trace Analysis

1. General

The recent general availability of solid state Ge(Li) gamma-ray detectors has made possible new applications of activation analysis to multi-element trace analysis. A simplified schematic representation of a Ge(Li) detector is given in Fig. 6. The principal advantage of these detectors is their excellent energy resolution for gamma-ray spectrometry [52]. While a typical $3'' \times 3''$ NaI(Tl) scintillation crystal may have a photopeak resolution of 50 KeV fwhm (*full width at half maximum*) for the [137]Cs

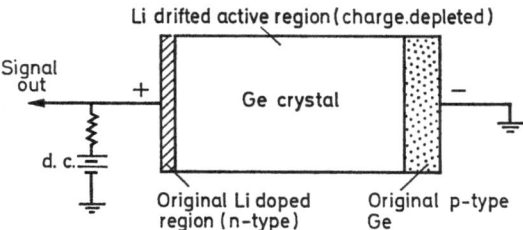

Fig. 6. Schematic representation of a Ge(Li) semiconductor type radiation detector

gamma-ray, a good Ge(Li) solid state detector will exhibit a resolution of 3—4 KeV fwhm. It is obvious that this high resolution will often permit resolution of very complex gamma-ray spectra without requiring complex computer spectrum resolution techniques.

Unfortunately, Ge(Li) detectors also have some disadvantages. Even the relatively large detectors of 35—40 cc active volume will have photopeak efficiencies of only approximately 5% that of a standard $3'' \times 3''$

NaI(Tl) scintillation detector for 1 MeV gamma-rays. The difference becomes even greater for higher energy gamma-rays, since the high atomic number of iodine in the NaI(Tl) detector results in a greater photoelectric efficiency at high energies. An additional disadvantage is that the Ge(Li) detector must be maintained at liquid nitrogen temperatures in order to maintain its properties. The need for a cryostat and for rather sophisticated noise-free amplification electronics makes the Ge(Li) detector system of 35—40 cc active volume at least 2 or 3 times more expensive than a simple $3'' \times 3''$ NaI(Tl) detector system.

The relatively low efficiency of solid state detectors presently available generally limits their use to activation techniques employing a nuclear reactor. The low flux and short half-life indicator radionuclides associated with 14 MeV neutron activation techniques in many cases do not permit accumulation of sufficient counts for good statistics with a Ge(Li) detector. Estimated sensitivities for potential non-destructive thermal neutron activation analyses using a large volume Ge(Li) detector are given in Fig. 7. Indicator radionuclides listed are all gamma-ray emitters, or decay by positron emission to yield annihilation radiation. Sensitivities are calculated on the basis of no spectral interferences and would be somewhat poorer in complex matrices. In many cases other indicator radionuclides than the one listed could be employed. The data presented should be regarded as merely representative of the sensitivity that might be obtained in more or less ideal situations.

Typically very few channels are included in a given photopeak in the analyzer display due to the high resolution of these detectors. Selection of the proper base-line for photopeak integration and errors due to chance dropping of data in an individual channel may still require dependence on computer programs to obtain accurate and consistent results [50]. The Compton edges in Ge(Li) spectra are also very sharp and care must be taken not to mistake the Compton edge from a higher energy gamma-ray interaction for a photopeak of analytical interest. The Compton edge of a higher energy gamma-ray interaction underlying a photopeak of interest may make precise photopeak integration extremely difficult. Adjustments of irradiation or decay periods may often be made to favor the radionuclide of interest. Even with the high resolution of the Ge(Li) detector some spectral interferences may dictate use of radiochemical separations or special techniques such as NaI(Tl)—Ge(Li) gamma-gamma coincidence spectrometry [51]. In the latter case the increased selectivity obtained may be at the expense of considerably reduced sensitivity.

In spite of the problems associated with the new solid state detectors there is no question that they are extremely useful in multi-element determinations and in many cases eliminate the need of time-consuming

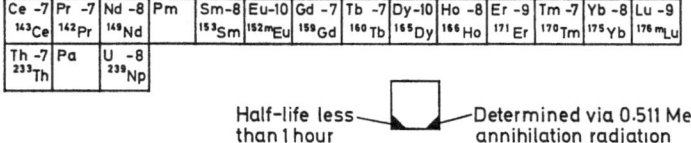

H																	He
Li	Be											B	C	N	O	F	Ne
Na −8 ^{24}Na	Mg −6 ^{27}Mg											Al −8 ^{28}Al	Si −4 ^{31}Si	P	S −3 ^{37}S	Cl −7 ^{38}Cl	Ar −8 ^{41}Ar
K −6 ^{42}K	Ca −5 ^{49}Ca	Sc −6 ^{46}Sc	Ti −7 ^{51}Ti	V −9 ^{52}V	Cr −5 ^{51}Cr	Mn −10 ^{56}Mn	Fe −3 ^{59}Fe	Co −7 ^{60m}Co	Ni −6 ^{65}Ni	Cu −8 ^{64}Cu	Zn −7 ^{69m}Zn	Ga −8 ^{72}Ga	Ge −7 ^{75}Ge	As −8 ^{76}As	Se −5 ^{75}Se	Br −8 ^{82}Br	Kr −7 ^{85m}Kr
Rb −5 ^{86}Rb	Sr −8 ^{87m}Sr	Y	Zr −6 ^{97}Zr	Nb −6 ^{94m}Nb	Mo −7 ^{99}Mo	Tc	Ru −7 ^{105}Ru	Rh −8 ^{104m}Rh	Pd −7 ^{109m}Pd	Ag −7 ^{108}Ag	Cd −7 ^{115}Cd	In −10 ^{116m}In	Sn −6 ^{125m}Sn	Sb −8 ^{122}Sb	Te −7 ^{131}Te	I −8 ^{128}I	Xe −7 ^{135m}Xe
Cs −8 ^{134}Cs	Ba −7 ^{139}Ba	La −8 ^{140}La	Hf −8 ^{180m}Hf	Ta −6 ^{182}Ta	W −8 ^{187}W	Re −9 ^{188}Re	Os −6 ^{193}Os	Ir −9 ^{194}Ir	Pt −8 ^{199}Pt	Au −9 ^{198}Au	Hg −8 ^{197}Hg	Tl	Pb	Bi	Po	At	Rn
Fr	Ra	Ac															

Ce −7 ^{143}Ce	Pr −7 ^{142}Pr	Nd −8 ^{149}Nd	Pm	Sm −8 ^{153}Sm	Eu −10 ^{152m}Eu	Gd −7 ^{159}Gd	Tb −7 ^{160}Tb	Dy −10 ^{165}Dy	Ho −8 ^{166}Ho	Er −9 ^{171}Er	Tm −7 ^{170}Tm	Yb −8 ^{175}Yb	Lu −9 ^{176m}Lu
Th −7 ^{233}Th	Pa	U −8 ^{239}Np											

Half-life less than 1 hour [□] Determined via 0.511 MeV annihilation radiation

Fig. 7. Estimated sensitivities for non-destructive thermal neutron activation analysis using a large volume Ge(Li) gamma-ray detector. The symbol Mn—10 signifies an approximate sensitivity of 10^{-10} grams for Mn using a reactor thermal neutron flux of 10^{13} neutrons cm^{-2} sec^{-1}. Irradiation times are to saturation activity or one hour, whichever is less. Calculations are based on counting the sample one centimeter from a 35 to 40 cc Ge(Li) detector (photopeak efficiency for ^{60}Co would be approximately 5% that for a 3″ × 3″ NaI(Tl) detector). Sensitivities listed are the amount of the element that would yield a gamma-ray photopeak counting rate of at least 10 counts/min for indicator radionuclides with half-lives greater than one hour and 100 counts/min for those with half-lives less than one hour. All indicator radionuclides listed have half-lives greater than 1 minute. In many cases other indicator radionuclides could be selected. For this compilation, based on an irradiation time of one hour or less, preference was given to shorter-lived radionuclides

radiochemical separations or use of complex computer processing of activation data. Some applications of these detectors in non-destructive activation analyses are discussed in the following section.

2. Some Applications

The utility of Ge(Li) detectors in activation analysis is best illustrated by their applications to trace element analyses in complex matrices such as *rocks, meteorites, and biological materials*. Immediately after irradiation of materials of these types the principal activities are due to ^{24}Na and ^{42}K, since sodium and potassium are major matrix components and have favorable activation properties. In order to determine trace element abundances without the necessity of employing radiochemical separations the samples are usually allowed to decay for at least five days, at which time these activities are appreciably reduced. Bremsstrahlung radiation from the beta decay of ^{32}P in biological materials having a high phosphorus content may also provide an interference to purely instrumental determinations. In the event it is desired to make use of short-lived indicator radionuclides in a determination, radiochemical separations for alkali elements and phosphorus may be required. Inorganic absorbers such as hydrated Sb_2O_5 for alkali elements and SnO_2 for phosphate ions have been employed for rapid elimination of these interferences in the analysis of biological materials [53].

The Ge(Li) gamma-ray spectrum of thermal neutron irradiated *cigarette tobacco* as obtained in our laboratory is shown in Fig. 8. In this case a 50 day decay period has eliminated the major portion of interferences from sodium, potassium and phosphorus activities. Assignments of photopeaks in this figure may be regarded as tentative, since half-lives of the individual peaks were not followed. As many as fifteen elements have been determined in tobacco products and biological standard kale

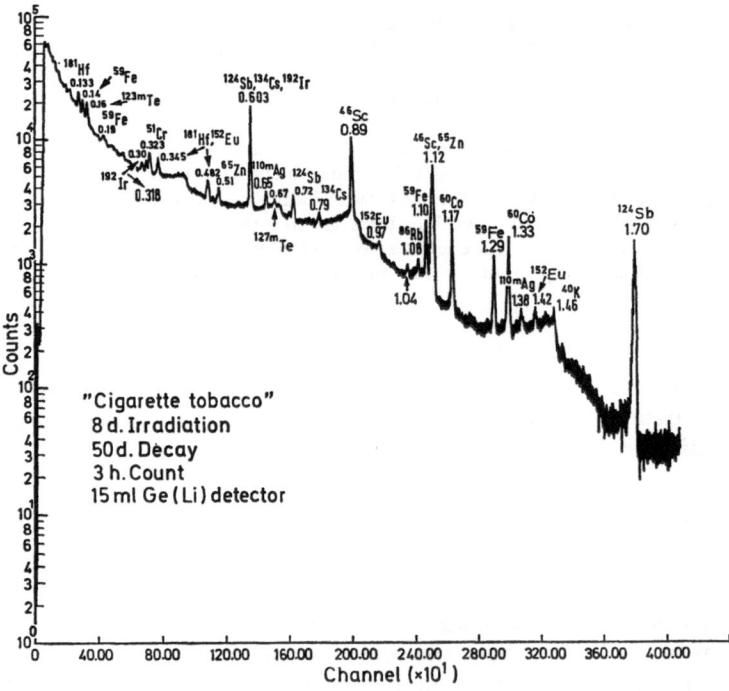

Fig. 8. Ge(Li) gamma-ray spectrum for thermal neutron irradiated cigarette tobacco

using a single reactor irradiation and purely instrumental techniques [54, 55]. The data obtained for the standard kale samples are in good agreement with data obtained by more time-consuming conventional analytical techniques Ge(Li) detectors have been used in the analysis of microplankton [56], human hair for forensic investigations [57], dental tissue [58], fish and animal tissues [59], barnacle shells [60] and a wide variety of other materials of biological origin.

The application of Ge(Li) detectors to the determination of *trace elements in rocks* is beautifully illustrated by the work of *Gordon* et al. [61] who were able to determine instrumentally 23 elements in a wide variety of igneous rocks. A useful discussion of sensitivities and potential interferences may also be found in this paper. The Ge(Li) gamma-ray spectrum of thermal neutron irradiated standard granite G–1 as obtained in our laboratory is shown in Fig. 9. Again, the tentative assignment of photopeaks is based

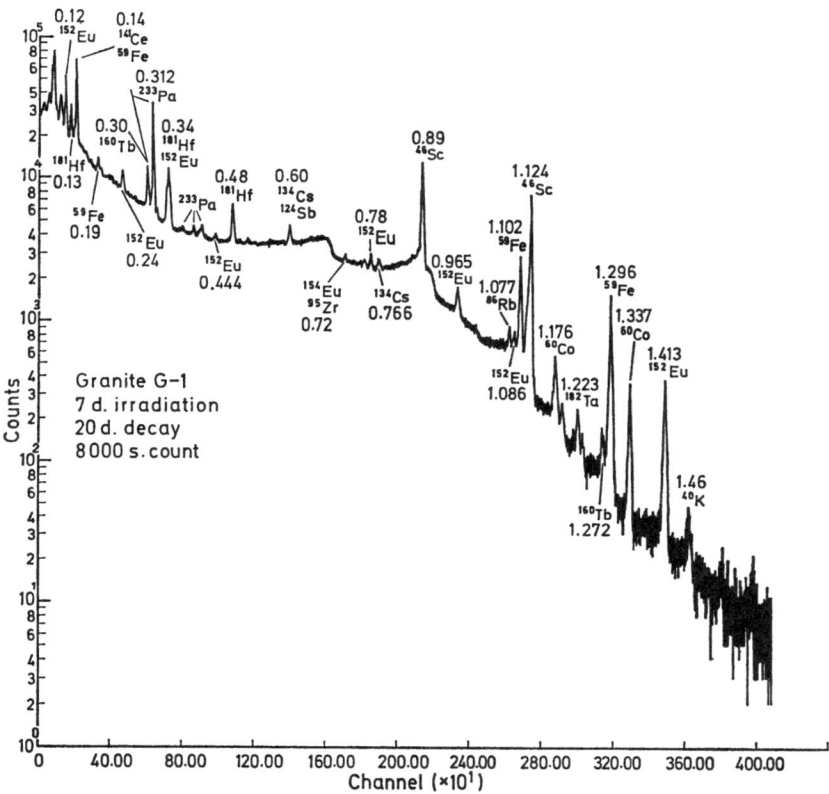

Fig. 9. Ge(Li) gamma-ray spectrum for thermal neutron irradiated standard granite rock, G—1

principally on energy calibrations. The period of decay for this sample was long enough that activities from sodium and potassium in the sample have decayed. The phosphorus content of this rock is negligible, as compared to the situation frequently encountered with biological materials. A somewhat different distribution of activities is observed in an irradiated tektite, as shown in Fig. 10. Due to the high background in the lower energy regions of Ge(Li) spectra (due principally to the Compton distributions from higher energy gamma-ray interactions), rapid radio-chemical group separations have been found to be desirable for the determination of certain rare earths, rubidium, cesium, strontium and barium in rocks [62].

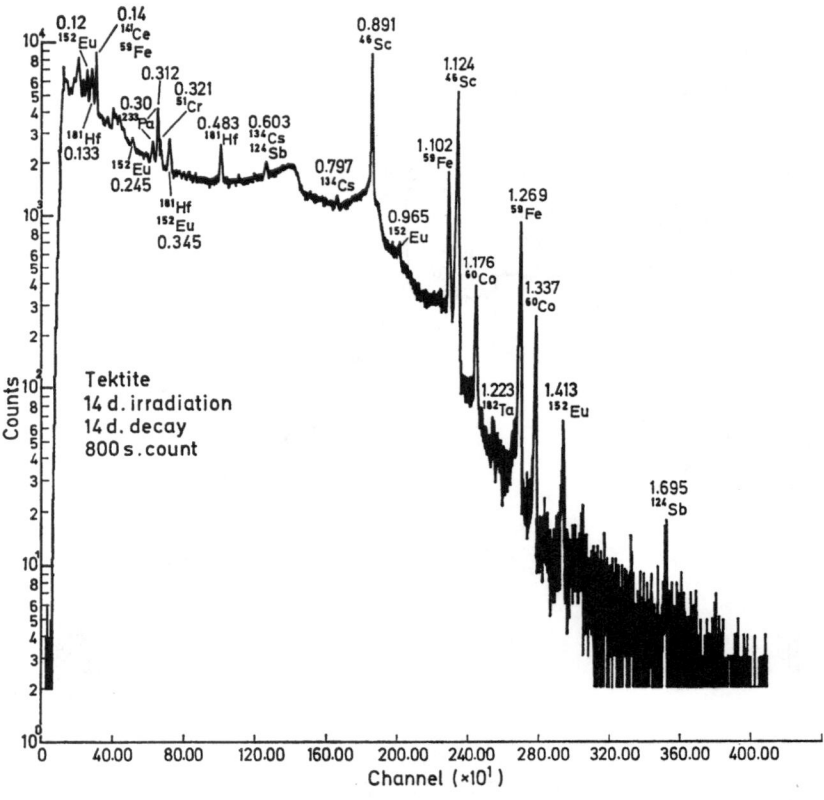

Fig. 10. Ge(Li) gamma-ray spectrum for a thermal neutron irradiated tektite. A major portion of the ^{124}Sb activity in this spectrum is due to antimony contamination in the quartz irradiation vial

Many applications which use Ge(Li) spectra of irradiated samples for identification purposes as a type of *"chemical fingerprint"* have recently appeared in the literature. The potential of this technique in the fields of forensics, archeology, and detection of art forgeries is just beginning to be explored.

C. Applications of Gamma-Gamma Coincidence Spectrometry

1. General

Gamma-gamma coincidence spectrometry has recently been applied in a variety of special cases to provide a high degree of resolution for radionuclides undergoing decay by cascade gamma-ray emission, or positron decay. In certain cases this technique may exhibit a greater freedom from interferences and higher counting efficiencies than can be obtained by currently available large volume Ge(Li) detectors.

A simplified schematic diagram of a *gamma-gamma coincidence system* is given in Fig. 11. The system illustrated consists of two 3″ × 3″ NaI(Tl) detectors coupled to photomultiplier tubes, two double delay line amplifiers, two single channel jitter-free analyzers to select the energy windows of interest, a fast coincidence module, and a multichannel analyzer for data accumulation. The photomultipliers are provided with voltage divider networks designed to reduce the gain shifts of the detector. A pulse generated in detector 1 and of energy within the window selected by single channel analyzer 1 and a pulse generated in detector 2 and of

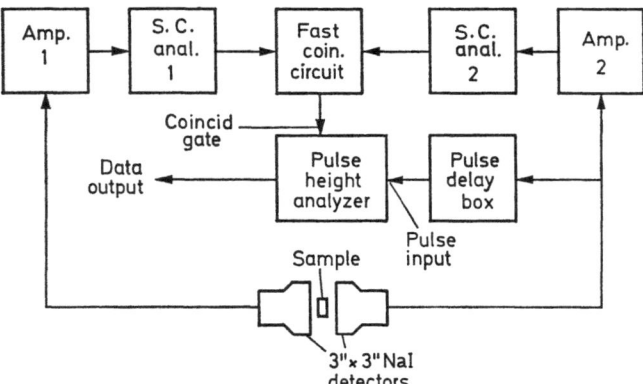

Fig. 11. Schematic diagram of a fast gamma-gamma coincidence spectrometer system

an energy selected by single channel analyzer 2 when in coincidence, will cause the fast coincidence unit to gate the multichannel analyzer through its slow coincidence gate input. The pulse from detector 2, simultaneously delayed in time by the variable delay box, is then accepted by the multichannel pulse height analyzer and stored in its memory. Placing the discrimination electronics before the fast coincidence module may reduce somewhat the time resolution of the system, but is effective in reducing the high Compton chance coincidence background in the low energy region of the spectrum. The system described as used in our laboratory [63,64] was found to have a coincidence resolution time of approximately 30 nanoseconds (full width at half maximum of a ^{60}Co variable delay curve). In order to achieve optimum precision and accuracy it is necessary to minimize long and short term gain shifts in the detector and amplifier circuits. Gain shifts which are a function of sample activity may be avoided by preparation of standards which will yield gross activities similar to that expected for the sample.

One simple approach to the preparation of appropriate standards where the gross activity of the sample may be large with respect to the activity of interest is to use the method of standard addition. In this technique a number of aliquants of the sample are spiked with varying amounts of the element to be determined and irradiated along with a similar sized unspiked sample. The specific activities of the spiked and unspiked samples are then plotted vs. the weight of added element as illustrated in Fig. 12. Extrapolation of this curve to zero specific activity will easily permit calculation of the unknown amount (x) of the element

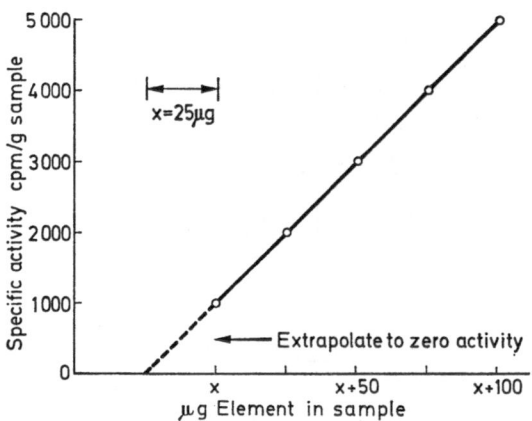

Fig. 12. Preparation of standards for non-destructive activation analysis using the method of standard addition

in the sample. This method also provides a degree of compensation for self-absorption effects in cases where it is necessary to use large samples.

Gamma-gamma coincidence techniques may be applied in either thermal or 14 MeV neutron activation analysis. However, the lower efficiency of the coincidence technique ($\leq 10\%$ with $3'' \times 3''$ NaI crystals) as compared to singles counting limits its use somewhat in the 14 MeV method, due to the lower flux and hence lower activity levels produced by commercial neutron generators. Its greatest utility in 14 MeV neutron activation is perhaps the measurement of 0.511 MeV annihilation radiation resulting from positron emitters produced by $(n, 2n)$ reactions. *Cuypers* and *Cuypers* [25] have tabulated the elements for which the 0.511 MeV photopeak is one of the major spectral features following activation with 14 MeV neutrons. A summary of their compilation is given in Table 1. Calculated sensitivities for 14 MeV activation followed by coincidence counting of annihilation radiation have been published by *Schulze* [65].

Table 1. *Elements which after 14 MeV neutron irradiation exhibit prominent 0.511 MeV positron annihilation photopeaks* [25]

Element determined	Indicator radionuclide	Half-life
Ag	^{106}Ag	24.0 m
Br	^{78}Br	6.5 m
Cl	^{34m}Cl	32.0 m
Cr	^{49}Cr	41.9 m
Cu	^{62}Cu	9.76 m
F	^{18}F	110. m
Fe	^{53}Fe	8.51 m
Ga	^{68}Ga	68.3 m
K	^{38}K	7.71 m
Mo	^{91}Mo	15.5 m
N	^{13}N	9.96 m
Ni	^{57}Ni	36.0 h
P	^{30}P	2.50 m
Pd	^{101}Pd	8.4 h
Pr	^{140}Pr	3.39 m
Sb	^{120}Sb	15.9 m
Sm	^{143}Sm	9.0 m
Zn	^{63}Zn	38.4 m

In the case of thermal neutron reactor irradiated samples relatively few positron emitters of analytical utility are produced. The indicator radionuclides ^{64}Cu (12.8 hour half-life) and ^{65}Zn (245 day half-life) produced from thermal neutron irradiation of copper and zinc, respectively, are the most promising for analytical work. Many indicator radionuclides produced by thermal neutron (n, γ) reactions, however, decay via cascade gamma-ray emission and may be selectively determined by coincidence techniques. A compilation of thermal neutron activation sensitivities using gamma-gamma coincidence has been prepared by *Wing* and *Wahlgren* [66]. Table 2 lists a few of the most promising elements which may be determined by means of cascade gamma-ray coincidences. This selection is based on radionuclides produced in high yield by thermal neutron activation and which have coincident gamma-rays which are prominent features of their spectrum. In special cases other indicator radionuclides might also be selected.

Table 2. *Thermal neutron activation determinations using gamma-gamma coincidence counting techniques*

Element determined	Indicator radionuclide	Half-life		Prominent coincident gamma-rays (MeV)
Ag	^{110m}Ag	260.	d	0.657, 0.884
As	^{76}As	26.5	h	0.559, 0.657
Au	^{198}Au	2.70	d	0.412, 0.676
Ba	^{131}Ba	12.	d	0.124, 0.496
Br	^{82}Br	35.3	h	0.554, 0.618
Cd	^{111m}Cd	48.6	m	0.150, 0.246
Ce	^{143}Ce	33.	h	0.232, 0.493
Cl	^{38}Cl	37.3	m	1.60, 2.17
Co	^{60}Co	5.26	y	1.17, 1.33
Cs	^{134}Cs	2.05	y	0.605, 0.796
Cu	^{64}Cu	12.8	h	0.511 annih. rad.
Dy	^{165}Dy	2.32	h	0.362, 0.633
Er	^{171}Er	7.52	h	0.112, 0.308
Eu	^{152m}Eu	9.3	h	0.122, 0.842
Ga	^{72}Ga	14.1	h	0.834, 2.20
Gd	^{161}Gd	3.7	m	0.102, 0.315
Ge	^{77}Ge	11.3	h	0.215, 0.417
Hf	^{181}Hf	42.5	d	0.133, 0.482
Hg	^{197m}Hg	24.	h	0.134, 0.165
Ho	^{166}Ho	26.9	h	0.080, 1.38
I	^{128}I	25.0	m	0.441, 0.528
In	^{116m}In	54.0	m	1.09, 1.29

Table 2 (continued)

Element determined	Indicator radionuclide	Half-life	Prominent coincident gamma-rays (MeV)
Ir	^{192}Ir	74.2 d	0.296, 0.316, others
La	^{140}La	40.2 h	0.329, 1.60
Lu	^{177}Lu	6.74 d	0.113, 0.208
Mn	^{56}Mn	2.58 h	0.847, 1.81
Mo	^{101}Mo	14.6 m	0.51, 1.56
Na	^{24}Na	15.0 h	1.37, 2.75
Nd	^{149}Nd	1.8 h	0.114, 0.424
Ni	^{65}Ni	2.56 h	0.368, 1.11
Os	^{193}Os	31. h	0.139, 0.322
Pt	^{197}Pt	18. h	0.077, 0.191
Rb	^{88}Rb	17.8 m	0.91, 1.85
Re	^{188}Re	16.7 h	0.155, 0.478
Rh	^{104m}Rh	4.41 m	0.051, 0.077
Ru	^{105}Ru	4.44 h	0.315, 0.47
Sb	^{124}Sb	60. d	0.603, 1.69
Sc	^{46}Sc	83.9 d	0.889, 1.12
Se	^{75}Se	120. d	0.136, 0.265
Sm	^{153}Sm	47. h	0.0697, 0.103
Sn	^{117m}Sn	14. d	0.159, 0.162
Ta	^{182}Ta	115. d	0.0678, 1.12
Tb	^{160}Tb	72.1 d	0.298, 0.879
Te	^{131}Te	25. m	0.150, 0.453
Ti	^{51}Ti	5.8 m	0.320, 0.605
U	^{239}Np	2.35 d	0.106, 0.278
W	^{187}W	23.9 h	0.072, 0.134
Yb	^{175}Yb	101. h	0.114, 0.283
Zn	^{65}Zn	245. d	0.511 annih. rad.

Early analytical determinations using the coincidence counting technique often merely set the two single channel analyzer windows to bracket the two or more coincident gamma-ray energies of interest. Scalers were used to simply record all coincident events. When working with complex matrices, many interfering activities may be produced. Chance Compton event coincidences in such a simple system may produce a high background level which may vary from sample to sample. It is desirable, therefore to use a multichannel pulse height analyzer rather than a scaler to record the spectrum generated in one detector which is in coincidence with events occurring within the energy window determined

W. D. Ehmann

by the second detector and single channel analyzer. The resultant spectrum may then be resolved by standard baseline subtraction methods to eliminate contributions from chance Compton coincidence events in the detectors.

A simple illustration of the improvement in resolution possible with a gamma-gamma coincidence system is presented in Fig. 13. A mixed source of ^{22}Na and ^{60}Co was counted in both the singles mode (A), and the coincidence mode (B), to accumulate approximately the same number of baseline corrected counts under the 0.511 MeV photopeak. In the coincidence mode the gating single channel analyzer window was set to bracket the 0.511 MeV energy region. In the coincidence mode spectrum the 0.511 MeV annihilation radiation photopeak from ^{22}Na is seen to be well defined on top of a low coincidence background which is easily resolved by baseline determination techniques. Selection of a proper baseline would be much more difficult in the case of the singles spectrum, due to the appreciably higher Compton baseline.

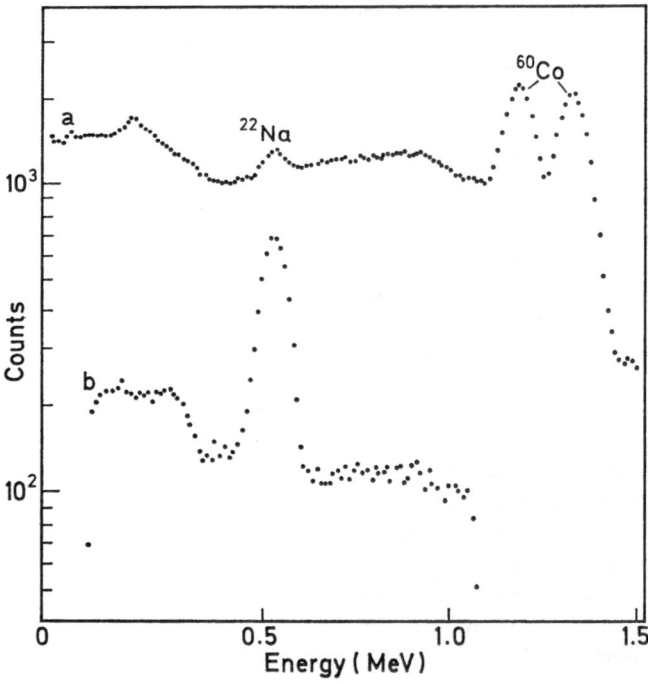

Fig. 13. Spectra of a mixed ^{22}Na-^{60}Co source in singles and coincidence counting modes illustrating improved baseline resolution (a = singles, b = coincidence)

Since the window setting in the second detector line is most often set to bracket the energy of only one of the coincident gamma-rays when their energy separation is large, the counting efficiency of such a system is low. Only half of the coincident events are recorded by the analyzer. A sum-coincidence spectrometer which permits almost twice the number of coincident events to be recorded has been described by *Hoogenboom* [67]. In this system a summing amplifier is used to add the two coincident pulses and the resultant pulse is fed to a single channel analyzer whose window is set to define the energy region of the summed pulse. This output is used to gate the multichannel analyzer together with the fast coincidence unit. If as determined by the coincidence unit, two events from the detectors are in coincidence, and if their summed energy falls within the window set by the single channel analyzer at the summer output, the event from the first detector is accumulated by the multichannel analyzer. Hence, both coincident gamma-ray peaks are recorded, resulting in an *improvement in the counting statistics* with no loss in resolution. In addition, deadtime effects in the multichannel analyzer are minimized since the high energy window setting of the single channel analyzer eliminates chance coincidence events that do not meet the sum energy requirements. Such a system has been used in the selective determination of antimony [68]. One disadvantage of such a system is that effects of baseline shifts are doubled in the use of the summing circuit.

In the use of the systems described above it is generally possible to determine only one indicator radionuclide during a single counting period. Setting narrow energy windows on the single channel analyzer(s) provides the desired selectivity and minimizes Compton interferences. A more elaborate coincidence system which permits multi-element determination while retaining the selectivity of the coincidence method has been described by *Perkins* and *Robertson* [69]. This system makes use of *multidimensional gamma-ray coincidence spectrometry*. The two primary NaI(Tl) detectors are surrounded by a large NaI(Tl) or plastic scintillator which is operated in anticoincidence to events recorded in the primary detectors. Compton scattering events in the primary detectors are therefore excluded from being recorded by the multichannel analyzer by detection of the scattered secondary photon by the external anticoincidence shield. When two gamma-rays are emitted in coincidence and are detected by photoelectric interactions in the two primary detectors, the event is recorded at a point in a matrix (often 64×64 channels) in the analyzer memory. The X and Y axes of the matrix correspond to the energy of gamma-rays received by the two primary detectors, respectively. Hence coincident gamma-rays of 1 MeV detected in detector No. 1 and 2 MeV detected in detector No. 2 would result in an event being stored in a memory position in the field of the matrix equivalent to 1 MeV on

77

the X axis and 2 MeV on the Y axis. While single channel analyzers may still be used to define the general energy field of interest, the window settings for both detectors may be set wide enough to permit determination of several coincident gamma-ray emitting indicator radionuclides simultaneously. Activities of individual radionuclides may then be resolved from the three dimensional array by integration of the volume of the peaks of interest, or by simple spectrum stripping operations. In effect, the multiparameter feature of the detection system permits simultaneous accumulation of data equivalent to a number of single channel analyzer window settings equal to the number of matrix elements. Fig. 14 illustrates the multiparameter display obtained in our laboratories [49] by coincidence counting of a thermal neutron irradiated stony meteorite. The spectrum represents the results of irradiating 0.23 grams of meteorite for one hour at a flux of $10^{13}\, n\, cm^{-2}\, sec^{-1}$ and counting for 40,000 seconds. The tops of the peaks correspond to approximately 20,000 counts. No anticoincidence Compton shield was used in this determination. The principal peaks are due to ^{46}Sc, ^{58}Co(from Ni), ^{60}Co, and ^{192}Ir.

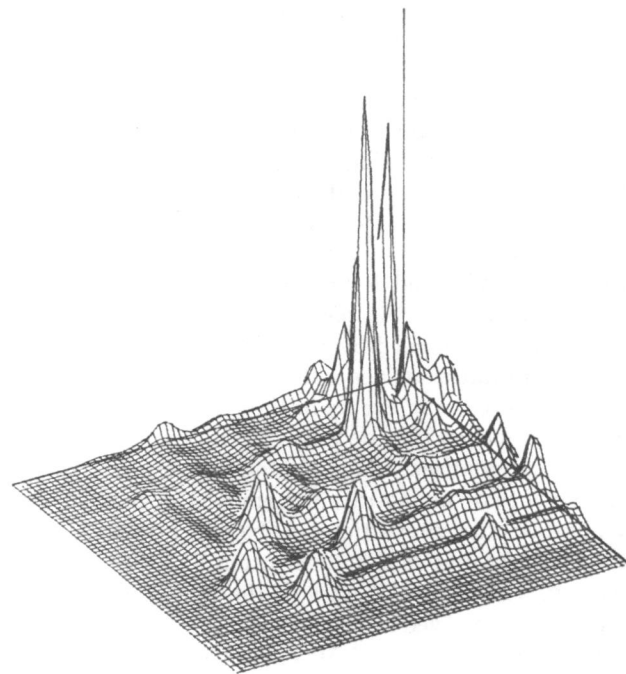

Fig. 14. Multiparameter coincidence spectrum of a thermal neutron irradiated stony meteorite. Energy windows set to cover range from approximately 0.2 to 2.0 MeV

Coincidence techniques have also been used for Compton interference reduction in the use of large volume Ge(Li) detectors together with plastic scintillator anticoincidence shields [70]. In some cases it might be desirable to use the coincidence electronics to gate the multichannel analyzer to accept only non-coincident pulses. In 14 MeV neutron activation procedures the annihilation radiation resulting from the decay of ^{13}N produced indirectly from the carbon in the plastic irradiation unit may be discriminated against by gating the analyzer to accept only non-coincident events.

2. Some Applications

A discussion of the coincidence technique with some general applications has been published by *Wahlgren, Wing* and *Hines* [71]. Many of the early applications of the technique made use of the fact that ^{64}Cu is one of the few radionuclides produced by thermal neutron irradiation for which the 0.511 MeV positron annihilation photopeak is a prominent feature of the spectrum. *Copper* has been determined in meteorites [72] and copper ores [73,74] by coincidence counting of ^{64}Cu annihilation radiation. The rapid and selective nature of the determination may have important applications in the on-line sorting of copper ores.

Greenland [75] has discussed the application of cascade gamma-ray coincidence techniques to the determination of *cesium and cobalt* in silicate rocks. *Bromine* has been determined in stony meteorites by this technique, using the cascade gamma-rays of [82]Br [76].

Herr and *Wolfle* [77] have described a coincidence method for the determination of trace amounts of *selenium and iridium* in minerals and several metals. The determination of iridium is an especially interesting problem, since this element is used as an indicator to measure the rate of infall of cosmic material on the earth. The indicator radionuclide, ^{192}Ir, produced by thermal neutron activation of iridium has a complex decay scheme with many cascade gamma-rays of potential analytical utility. The availability of a number of coincident pairs provides an additional opportunity to check for potential interferences when iridium is a minor component in complex matrices. The high thermal neutron cross section of [191]Ir and the application of coincidence counting makes it possible to determine iridium non-destructively in complex matrices at the submicrogram level, using irradiations of less than one hour at a flux of 10^{13} n cm^{-2} sec^{-1}. Fig. 15 illustrates the singles and coincidence spectra for a thermal neutron irradiated stony meteorite [64]. In the singles spectrum the photopeaks of ^{192}Ir are almost completely disguised by the presence of ^{51}Cr at approximately 0.32 MeV, annihilation radiation, and the Compton distribution from higher energy gamma-rays. The strong interference of ^{51}Cr is not removed by use of a Ge(Li) detector, although a determination may be made by use of one of the less abundant ^{192}Ir gamma-rays, if the appreciable decrease in sensitivity may be accommodated. The coincidence spectrum of the same meteorite illustrated in Fig. 15, clearly shows the spectrum of ^{192}Ir. The intense photopeak at approximately 0.30—0.32 MeV was shown by standard addition of ^{51}Cr and half-life to be entirely due to ^{192}Ir and may be used for the iridium determination. The detection efficiency by means of this coincidence method has been shown to be approximately 10 times greater than that for meteoritic iridium determinations using a 12 cc Ge(Li) detector.

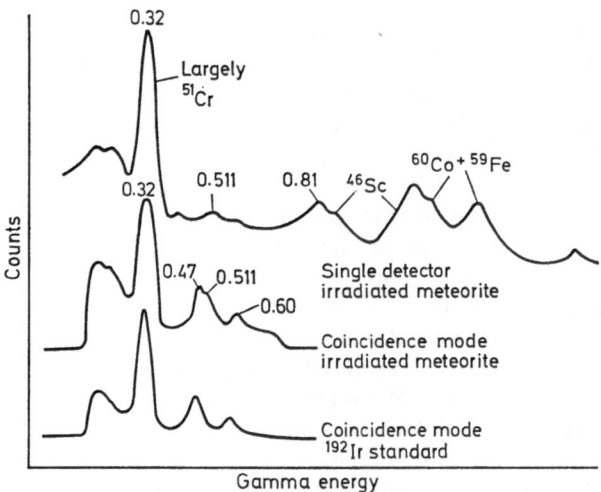

Fig. 15. Singles and coincidence spectra for a thermal neutron irradiated stony meteorite

Fujii et al. [78] have developed a rapid method for the determination of *praseodymium*, using 14 MeV neutron activation and gamma-gamma coincidence counting.

The coincidence technique has also been applied to the activation analysis of *biological materials* [79-82] and *forensic materials* [83].

One of the earlier and still one of the most interesting applications of coincidence counting in activation analysis is the isotopic determination of ^6Li by *Coleman* [84]. In this method aqueous solutions of LiOH are irradiated with thermal neutrons to produce tritons by means of the $^6Li(n,t)^4He$ reaction. The tritons then react with the oxygen in the aqueous solution by the $^{16}O(t,n)^{18}F$ reaction. The ^{18}F is a positron emitter with a half-life of 109.7 minutes and is detected by coincidence measurement of 0.511 MeV annihilation radiations. The method permits multiple isotopic determinations on the same sample. The procedure is rapid and has been shown to yield data in good agreement with those obtained by use of mass spectrometry, which is a destructive technique.

D. Other Non-Destructive Techniques

1. General

The use of 14 MeV neutron activation principally for major elements, Ge(Li) detectors for trace elements following thermal neutron irradiations, and gamma-gamma coincidence techniques for positron or cascade gamma-ray emitters as discussed in the previous sections, provide the analyst with powerful tools for devising schemes for non-destructive analysis. A few additional activation techniques which may be useful in special applications are discussed briefly below. In most of these cases rather sophisticated instrumentation is required. It is unlikely, therefore, that these techniques will be in routine use in a facility devoted principally to analytical applications. In some cases, however, arrangements may be made for part time use of a more extensive nuclear facility for a specific analytical problem.

2. Photoactivation Analysis

A number of lighter elements, such as carbon, nitrogen and oxygen may not be determined at the ppm level, by the usual neutron activation techniques. Activation with photons of 10 to 70 MeV generated by an electron linear accelerator or a betatron [85] makes it possible to determine these and many other elements at the microgram level, or below. The principal photon induced nuclear reactions in this energy region are the (γ, γ') and (γ, n) reactions. In the case of the (γ, n) reaction the product radionuclide is most often a positron emitter and additional selectivity may be obtained by coincidence counting of the annihilation radiations. The nuclear reactions $^{16}O(\gamma, n)^{15}O$ $H = 2.1$ minutes, $^{14}N(\gamma, n)^{13}N$ $H = 10$ minutes, and $^{12}C(\gamma, n)^{11}C$ $H = 20.5$ minutes all yield positron emitters. These elements are of considerable interest in the analysis of biological materials. Resolution of the activities from oxygen and carbon is easily accomplished by decay curve analysis, due to their greatly different half-lives. The presence of nitrogen in mixtures containing oxygen and/or carbon makes decay curve resolution much more difficult. The threshold energy for the $^{14}N(\gamma, n)^{13}N$ reaction is, however, considerably lower than that for the other two reactions and selective analyses for nitrogen may be accomplished with reduced sensitivity, using photons of approximately 12 MeV. The reduced sensitivity is due to the fact that the activation cross section for the reaction on nitrogen is extremely small for photon energies less than 20 MeV. For photon activation of matrices containing oxygen, carbon and nitrogen, radiochemical separation of nitrogen following irradiation is often found to be desirable. As with neutron activation analysis, primary interference reactions may be a problem. In some cases where the desired reaction and the interference reaction have appreciably different threshold energies (for example, $^{14}N(\gamma, n)^{13}N$ threshold = 10.6 MeV and $^{16}O(\gamma, t)^{13}N$ threshold = 25 MeV), discrimination may be simply accomplished by selection of the appropriate electron bombarding energy in the accelerator.

Many articles relating to the techniques and applications of photo-activation analysis have appeared in the literature in the last several years. Especially useful reviews have been published by *Albert* [86,87], *Engelmann* et al. [88], and *Lutz* [99].

3. Charged Particle Activation Analysis

Protons, deuterons, tritons, helium-3 ions, and alpha particles have all been used as activating particles for activation analysis. Cyclotrons capable of accelerating the charged particles to energies of approximately 15 MeV for singly charged particles, or 30 MeV for doubly charged particles are commonly used. For protons, (p, n), (p, pn), $(p, 2n)$, and (p, α) reactions are among the most common. The number of potentially useful reactions increases with increasing bombardment energy.

The method has several limitations. First, relatively few analytical chemists have routine access to accelerators of the type required. In addition, the relatively small penetration of charged particles of the above energy restricts the depth of activation in the sample to a layer often less than 1 mm thick. In some cases where only the oxygen content of a metal surface is desired, this may be an advantage [89]. The attenuation of the charged particles in the sample liberates an appreciable amount of heat which may lead to alteration of the sample unless it is adequately cooled. In this respect, the technique may not be regarded to be truly non-destructive.

The technique has been widely applied to analyses for lighter elements such as carbon, oxygen, sulphur, nitrogen and boron in high purity metals and semiconductor materials [86-88,90-92]. Neutron activation techniques for these elements do not provide sufficient sensitivity for many applications. In addition, the high neutron capture cross sections of many metals prevent the use of neutron activation techniques in the determination of trace impurities in the high purity metal matrix.

4. Prompt Gamma-Ray Analysis

The prompt gamma-rays emitted following neutron or charged particle interactions with the target nuclide may be used as a basis for non-destructive analyses. The important advantage of this technique is that the determination does not depend in any manner on the half-life of a product radionuclide. In fact, using this technique, the product nuclide need not even be radioactive. Many conventional activation determinations are limited in their sensitivities by short half-life product radionuclides, or the fact that the most abundant or highest cross section isotope of the element to be determined leads to a stable product on irradiation.

Following capture of a thermal neutron (the most common activating particle for this technique) the resultant compound nucleus promptly decays, usually through several intermediates, yielding a spectrum of gamma-rays characteristic of the nuclide. The physical setup of the irradiation and counting components is extremely important in minimizing interferences. Ordinarily a collimated beam of thermal neutrons is passed out of a suitable reactor port and allowed to impinge upon the sample which is in the form of a circular disc. The sample holder must be designed to minimize the mass of supporting material intersected by the beam. A NaI(Tl) or Ge(Li) detector with collimator and appropriate shielding is placed to aim at the sample at right angles to the neutron beam. The detector must be shielded from scattered neutrons as well as from spureous gamma-rays generated in the vicinity of the counting station. Detailed descriptions of such a setup have been given by *Greenwood* and *Reed* [93] and *Lombard* et al. [94]. A fast/slow sum-coincidence spectrometer has been used by *Lussie* and *Brownlee* [95] to further reduce the spureous gamma-ray background encountered in capture gamma-ray detection.

It is of interest to note that the method is capable of isotopic analysis of elements having adjacent stable isotopes and has also been used for the direct determination of hydrogen in plastics [94]. The prompt gamma-ray emission following proton or deuteron bombardment has been used to measure the carbon content of steels [96,97]. The method has also been widely used for the activation determination of boron. The principal disadvantage of the method is the requirement of direct access to a port in a nuclear reactor or a particle accelerator. Isotopic sources of neutrons have been used in some cases where high sensitivity is not required. The method should become more useful in the future when high intensity ^{252}Cf neutron sources become available. The neutron emission rate of a one curie ^{252}Cf spontaneous fission neutron source is 300 times that which may be obtained from any other one curie isotopic neutron source [98]. The use of these sources would reduce the shielding problems encountered in the use of nuclear reactors for prompt-gamma analysis. Unfortunately, it is certain to be many years before these sources are widely available.

5. Neutron Activation Followed by Delayed Neutron Counting

Neutron irradiation of fissionable nuclides results in the production of many fission product radionuclides which decay by negatron emission. In some cases, the negatron decay of the fission product radionuclide leads to an excited state of a daugther radionuclide which promptly emits a neutron to attain a "magic number" neutron configuration. An example is the fission product $^{87}Br(H = 55$ sec.) which decays by negatron emission to an excited state of ^{87}Kr which in turn promptly emits a neutron to form stable ^{86}Kr with the magic neutron number 50. Since the emission of the neutron from ^{87}Kr is prompt, the rate of neutron emission from the irradiated sample decays with the half-life of the delayed neutron emitter precursor, ^{87}Br.

The fact that neutrons can be detected with reasonably high efficiency and with minimal interferences from other radiations permits the practical determination of fissionable species such as isotopes of uranium and thorium by delayed neutron counting. The known delayed neutron emitter precursors are all short lived and the irradiated samples are counted with $^{10}BF_3$-filled proportional counters immediately after irradiation without any separation chemistry.

In early work gross counting of delayed neutrons was used to determine the abundance of a single fissionable nuclide known to be in the sample. *Brownlee* [101] has reported techniques by which two or more fissionable species may be determined at the submicrogram level in a single irradiated sample. Nuclides fissionable only with fast neutrons may also be determined by this technique. One of the more interesting applications of the method is in the non-destructive determination of uranium and thorium at trace levels in minerals, rocks, and stony meteorites [102,103].

III. Summary

It has been the purpose of this paper to review some of the more recent developments in the use of activation techniques for the essentially non-destructive determination of elemental or isotopic abundances. The term non-destructive is used in the sense that the gross chemical and physical properties of the sample remain essentially unaltered. In the case of reactor irradiations for long periods of time, the samples will retain an appreciable level of radioactivity which may limit their use in certain later applications. Long exposure to radiation will also result in the degradation of certain biological materials. The essential point is that there are now a wide variety of activation techniques that will yield abundance data rapidly and accurately and still permit retention of the sample for other studies, forensic evidence, or museum display.

For ultimate sensitivity, activation techniques employing radio-chemical separations must still be employed. Chemical separations using inactive carriers may be used to obtain radiochemically pure samples that may be counted with beta particle detectors which have much higher counting efficiencies than any gamma-ray detector commonly used. In many cases a rapid group separation (as in the separation of the alkali group elements from irradiated biological materials) will permit the rapid determination of many more elements than can be determined by strictly instrumental means.

The current availability of small portable 14 MeV neutron generators and the future availability of high intensity ^{252}Cf spontaneous fission neutron sources will certainly result in the wide spread use of activation techniques for non-destructive "on-stream" product analysis in industry. The cost of the required instrumentation for many types of activation analysis is not excessive, as compared to the cost of other modern analytical instrumentation. The simple "off-on" operation of the new sealed-tube neutron generators and minimal maintenance associated with the use of an isotopic ^{252}Cf neutron source will permit operation of the analytical facility with technician-level personnel. The versatility of the activation technique justifies its inclusion among the other major analytical techniques employed in any modern analytical facility.

Acknowledgement: Some of the work described in this paper was supported in part by U.S. Atomic Energy Commision Contract AT-(40-1)-2670.

IV. References

[1] *Hevesy, G.,* and *H. Levi:* The Action of Neutrons on the Rare Earth Elements. Mathematisk-Fysiske Meddelelser *14,* 3 (1936).
[2] *Lutz, G. J., R. J. Boreni, R. S. Maddock,* and *W. W. Meinke:* Activation Analysis: A Bibliography, Parts I and II. National Bureau of Standards Technical Note 467. Washington, D. C.: U.S. Government Printing Office 1968.
[3] *Bowen, H. J. M.,* and *D. Gibbons:* Radioactivation Analysis. Oxford: Clarendon Press 1963.
[4] *Cali, J. P. (Ed.):* Trace Analysis of Semiconductor Materials. Oxford: Pergamon Press 1964.
[5] Centre D'Etudes Nucleaires De Saclay: L'Analyse Par Radioactivation Et Ses Application Aux Sciences Biologiques — 3e Colloque International De Biologie De Saclay. Paris: Presses Universitaires De France 1963.

W. D. Ehmann

6) *Crouthamel, C. E.* (Ed.): Applied Gamma-Ray Spectrometry. Oxford: Pergamon Press 1960.

7) *Guinn, V. P., D. Lichtman, R. B. McQuistan, and J. E. Strain (H. A. Elion and D. C. Stewart*, Eds.): Analytical Chemistry, Series IX, Volume 4, Parts 2 and 3. Oxford: Pergamon Press 1965.

8) International Atomic Energy Agency: Joint Commission on Applied Radioactivity (I. C. S. U.): Radioactivation Analysis-Proceedings of the Radioactivation Analysis Symposium held in Vienna, Austria, June 1959. London: Butterworth 1960.

9) International Atomic Energy Agency: Radiochemical Methods of Analysis-Proceedings of the Symposium on Radiochemical Methods of Analysis held in Salzburg, Austria, October, 1964. Vienna: International Atomic Energy Agency 1965.

10) *Koch, R, C.:* Activation Analysis Handbook. New York: Academic Press 1960.

11) *Lenihan, J. M. A.*, and *S. J. Thomson* (Eds.): Activation Analysis-Principles and Applications. New York: Academic Press 1965.

12) *Lyon, W. S.* (Ed.): Guide to Activation Analysis. Princeton, N. J.: D. Van Nostrand Company Inc. 1964.

13) — Modern Trends in Activation Analysis—Proceedings of the 1961 International Conference in College Station, Texas, December, 1961. A and M College of Texas, College Station, Texas (1961).

14) — Modern Trends in Activation Analysis—Proceedings of the 1965 International Conference in College Station, Texas, April, 1965. A and M College of Texas, College Station, Texas (1965).

15) — Modern Trends in Activation Analysis — Proceedings of the 1968 International Conference in Gaithersburg, Maryland, October, 1968. National Bureau of Standards, Washington, D. C. (1968).

16) *Moses, A. J.:* Nuclear Techniques in Analytical Chemistry. Oxford: Pergamon Press 1964.

17) *Taylor, D.:* Neutron Irradiation and Activation Analysis. Princeton, N. J.: D. Van Nostrand Company, Inc. 1964.

18) *Vogt, J. R., W. D. Ehmann*, and *M. T. McEllistrem:* An Automated System for Rapid and Precise Fast Neutron Activation Analysis. Intern. J. Appl. Radiation Isotopes *16*, 573 (1965).

19) *Schmitt, R. A., R. H. Smith*, and *G. G. Goles:* Abundances of Na, Sc, Cr, Mn, Fe, Co, and Cu in 218 Individual Chondrules via Activation Analysis, 1. J. Geophys. Res. *70*, 2419 (1965).

20) *Coleman, R. F.*, and *J. L. Perkin:* The Determination of the Oxygen Content of Beryllium Metal by Activation. Analyst *84*, 233 (1959).

21) *Volborth, A.*, and *H. E. Banta:* Oxygen Determination in Rocks, Minerals, and Water by Neutron Activation. Anal. Chem. *35*, 2203 (1963).

22) *Wing, J.:* Simultaneous Determination of Oxygen and Silicon in Meteorites and Rocks by Non-destructive Activation Analysis. Anal. Chem. *36*, 559 (1964).

23) *Vogt, J. R.*, and *W. D. Ehmann:* An Automated Procedure for the Determination of Oxygen Using Fast Neutron Activation Analysis; Oxygen in Stony Meteorites. Radiochim. Acta *4*, 24 (1965).

24) *Volborth, A.:* Precise and Accurate Oxygen Determination by Fast Neutron Activation. Fortschr. Mineral. *43*, 10 (1966).

25) *Cuypers, M.*, and *J. Cuypers:* Gamma-Ray Spectra and Sensitivities for 14 MeV Neutron Activation Analysis. Activation Analysis Research Laboratory, Texas A and M University, College Station, Texas (1966).

26) *Anders, O. U.*, and *D. W. Briden:* A Rapid, Non-destructive Method of Precision Oxygen Analysis by Neutron Activation. Anal. Chem. *36*, 287 (1964).

27) *Cockcroft, J. D.*, and *E. T. S. Walton:* Experiments with High Velocity Positive Ions. I. Further Developments in the Method of Obtaining High Velocity Positive Ions. Proc. Roy. Soc. (London) *A 136*, 619 (1932).

28) *Moak, C. D., H. Reese*, and *W. M. Good:* Design and Operation of a Radio-Frequency Ion Source for Particle Accelerators. Nucleonics *9*, No. 3, 18 (1951).

29) —, *H. E. Banta, J. N. Thurston, J. W. Johnson*, and *R. F. King:* Duo Plasmatron Ion Source for Use in Accelerators. Rev. Sci. Instr. *30*, 694 (1959).

30) *Johnson, A. G.:* Tritium Considerations Associated with the Operation of Cockcroft-Walton-Type Neutron Generators. U. S. Atomic Energy Commision. Unpublished report (1965).

31) *Jessen, P. L.:* Long Term Operating Experience with High Yield Sealed Tube Neutron Generators. The 1968 International Conference Modern Trends in Activation Analysis, Gaithersburg, Maryland, October 7—11, 1968, Paper 159.

32) *Downton, D. W.*, and *J. D. L. H. Wood:* A 10^{11} Neutrons per Second Tube for Activation Analysis. The 1968 International Conference Modern Trends in Activation Analysis, Gaithersburg, Maryland, October 7—11, 1968, Paper 25.

33) *Reifenschweiler, O.:* A High Output Sealed-Off Neutron Tube with High Reliability and Long Life. The 1968 International Conference Modern Trends in Activation Analysis, Gaithersburg, Maryland, October 7—11, 1968, Paper 20.

34) *Meinke, W. W.:* Pneumatic Tubes Speed Activation Analysis. Nucleonics *17*, No. 9, 86 (1959).

35) *Steel, E. L.*, and *W. W. Meinke:* Determination of Oxygen by Activation Analysis with Fast Neutrons Using a Low-Cost Portable Neutron Generator. Anal. Chem. *34*, 185 (1962).

36) *Strain, J. E.:* Use of Neutron Generators in Activation Analysis. Progress in Nuclear Energy, Series IX, Volume 4, Part 3, pp. 137—157. Oxford: Pergamon Press 1965.

37) *Vogt, J. R.:* Accelerator Systems for Activation Analysis — A Comparative Survey. Developments in Applied Spectroscopy, Vol. 6, pp. 161—176. New York: Plenum Press 1968.

38) *Ehmann, W. D.*, and *D. M. McKown:* Heat-Sealed Polyethylene Sample Containers for Neutron Activation Analysis. Anal. Chem. *40*, 1758 (1968).

39) *Dyer, F. F., L. C. Bate*, and *J. E. Strain:* Three-Dimensionally Rotating Sample Holder for 14-Million Electron Volt Neutron Irradiations. Anal. Chem. *39*, 1907 (1967).

40) *Priest, G. L., F. C. Burns*, and *H. F. Priest:* Uniform Neutron Irradiation of Inhomogeneous Samples. Anal. Chem. *39*, 110 (1967).

41) *Broadhead, K. G.*, and *D. E. Shanks:* The Application of 2.8-MeV (D,d) Neutrons to Activation Analysis. Intern. J. Appl. Radiation Isotopes *18*, 279 (1967).

42) *Mathur, S. C.*, and *G. Oldham:* Interferences Encountered in 14 MeV Neutron Activation Analysis. Nucl. Energy, Sept.-Oct., pp. 136—141 (1967).

43) *Mott, W. E.*, and *J. M. Orange:* Precision Analysis with 14 MeV Neutrons. Anal. Chem. *37*, 1338 (1965).

44) *Vogt, J. R.*, and *W. D. Ehmann:* Silicon Abundances in Stony Meteorites by Fast Neutron Activation Analysis. Geochim. Cosmochim. Acta *29*, 373 (1965).

45) *Broadhead, K. G., D. E. Shanks*, and *H. H. Heady:* Fast-Neutron Activation Analysis in Molten Salt Electrometallurgical Research. Modern Trends in Activation Analysis-Proceedings of the 1965 International Conference in College Station, Texas, April, 1965. A and M College of Texas, College Station, Texas, pp. 39—43 (1965).

46) *Wood, D. E., P. L. Jessen,* and *R. E. Wood:* Industrial Application of Fast Neutron Activation Analysis for Protein Content of Food Products. Paper presented at the 52nd Annual Meeting of the American Association of Cereal Chemists, Los Angeles, California, April, 1967.

47) *Guinn, V. P., B. J. Kleinstein,* and *G. C. Mull:* The Determination of Oxygen-18 by Activation Analysis with 14 MeV Neutrons and the Oxygen-18 (n,α) Carbon-15 Reaction. Trans. Am. Nucl. Soc. *9,* 83 (1966).

48) *Guinn, V. P.:* Activation Analysis with Particular Attention to the Detection of Stable Tracers. In: Isotopes in Experimental Pharmacology (*L. J. Roth,* Ed.). Chicago: University of Chicago Press 1965.

49) *Morgan, J. W.:* Unpublished work. Lexington: University of Kentucky, 1969.

50) *Ralston, H. R.,* and *G. E. Wilcox:* A Computer Method of Peak Area Determination from Ge(Li) Gamma Spectra. The 1968 International Conference Modern Trends in Activation Analysis, Gaithersburg, Maryland, October 7—11, 1968, Paper 37.

51) *Currie, R. L., R. McPherson,* and *G. H. Morrison:* A Coindence-Anticoincidence System for Activation Analysis Employing a Split (NaI(Tl)) Annulus and a Large Volume Ge(Li) Detector. The 1968 International Conference Modern Trends in Activation Analysis, Gaithersburg, Maryland, October 7—11, 1968, Paper 131.

52) *Voight, A. F., D. E. Becknell,* and *Sr. L. Menapace:* Comparison of Solid State and Scintillation Gamma-Ray Spectrometry in Analysis. The 1968 International Conference Modern Trends in Activation Analysis, Gaithersburg, Maryland, October 7—11, 1968, Paper 106.

53) *Meloni, S., A. Brandone,* and *V. Maxia:* Chromium Separation by Inorganic Exchangers in Activation Analysis of Biological Materials. The 1968 International Conference Modern Trends in Activation Analysis, Gaithersburg, Maryland, October 7—11, 1968, Paper 52.

54) *Nadkarni, R. A.,* and *W. D. Ehmann:* Instrumental Neutron Activation Analysis of Tobacco Products. The 1968 International Conference Modern Trends in Activation Analysis, Gaithersburg, Maryland, October 7—11, 1968, Paper 47.

55) — — Determination of Trace Elements in Biological Standard Kale by Neutron Activation Analysis. J. Radioanal. Chem., in press (1969).

56) *Merlini, M., O. Ravera,* and *C. Bigliocca:* Non-destructive Determination of Elements in Specific Freshwater Microplankton by Activation Analysis. The 1968 International Conference Modern Trends in Activation Analysis, Gaithersburg, Maryland, October 7—11, 1968, Paper 167.

57) *Perkins, A. K.,* and *R. E. Jervis:* Recent Forensic Applications of Instrumental Activation Analysis. The 1968 International Conference Modern Trends in Activation Analysis, Gaithersburg, Maryland, October 7—11, 1968, Paper 60.

58) *Nadkarni, R. A., D. E. Flieder,* and *W. D. Ehmann:* Instrumental Neutron Activation Analysis of Biological Materials. Radiochim. Acta, *11,* 97 (1969).

59) *Rancitelli, L. A., J. A. Cooper,* and *R. W. Perkins:* The Multielement Analysis of Biological Material by Neutron Activation and Direct Instrumental Techniques. The 1968 International Conference Modern Trends in Activation Analysis Gaithersburg, Maryland, October 7—11, 1968, Paper 136.

60) *Gordon, C. M.,* and *R. E. Larsen:* Neutron Activation Analysis of Barnacle Shells. The 1968 International Conference Modern Trends in Activation Analysis, Gaithersburg, Maryland, October 7—11, 1968, Paper 29.

61) *Gordon, G. E., K. Randle, G. G. Goles, J. B. Corliss, M. H. Beeson*, and *S. S. Oxley:* Instrumental Activation Analysis of Standard Rocks with High-Resolution γ-ray Detectors. Geochim. Cosmochim. Acta 32, 369 (1968).

62) *Higuchi, H., K. Tomura, H. Takahashi, N. Onuma*, and *H. Hamaguchi:* Use of a Ge(Li) Detector After Simple Chemical Group Separation in the Activation Analysis of Rock Samples. IV. Simultaneous Determination of Strontium and Barium. The 1968 International Conference Modern Trends in Activation Analysis, Gaithersburg, Maryland, October 7—11, 1968, Paper 48.

63) *Ehmann, W. D.*, and *D. M. McKown:* The Nondestructive Determination of Iridium in Meteorites Using Gamma-Gamma Coincidence Spectrometry. The 1968 International Conference Modern Trends in Activation Analysis, Gaithersburg, Maryland, October 7—11, 1968, Paper 46.

64) — — Instrumental Activation Analysis of Meteorites Using Gamma-Gamma. Coincidence Spectrometry. Anal. Letters 2 (1), 49 (1969).

65) *Schulze, W.:* Application of Coincidence Methods in Activation Analysis. Modern Trends in Activation Analysis-Proceedings of the 1965 International Conference in College Station, Texas, April, 1965. A and M College of Texas, College Station Texas, pp. 272—278 (1965).

66) *Wing, J.*, and *M. A. Wahlgren:* Detection Sensitivities in Thermal-Neutron Activation. Argonne National Laboratory Report 6953 (1965).

67) *Hoogenboom, A. M.:* A New Method in Gamma-Ray Spectroscopy: A Two Crystal Scintillation Spectrometer with Improved Resolution. Nucl. Instr. 3, 57 (1958).

68) *Adams, F.*, and *J. Hoste:* Activation Analysis of Antimony by Sum-Coincidence Spectrometry. Nucleonics 22, No. 3, 55 (1964).

69) *Perkins, R. W.*, and *D. E. Robertson:* Selective and Sensitive Analysis of Activation Products by Multidimensional Gamma-Ray Spectrometry. Modern Trends in Activation Analysis-Proceedings of the 1965 International Conference in College Station, Texas, April, 1965. A and M College of Texas, College Station, Texas, pp. 48—57 (1965).

70) *Cooper, R. D.*, and *G. L. Brownell:* A Large Coaxial Ge(Li) Detector with Plastic Anticoincidence Scintillator for Activation Analysis. Nucl. Instr. Methods 51, No. 1, 72 (1967).

71) *Wahlgren, M., J. Wing*, and *J. Hines:* A Fast-Sum Coincidence Spectrometer and Sensitivity Compilation for Activation Analysis. Modern Trends in Activation Analysis—Proceedings of the 1965 International Conference in College Station, Texas, April, 1965. A and M College of Texas, College Station, Texas, pp. 134—139 (1965).

72) *Schmitt, R. A.*, and *R. H. Smith:* Research on Elemental Abundances in Meteoritic and Terrestrial Matter. General Atomics Report 6642 (1965).

73) *Ramdohr, H. F.:* Activation Analysis for Copper Ore Sorting. Kerntechnik 5, 204 (1963).

74) *Gorski, L., W. Kusch*, and *J. Wojtkowska:* Fast Neutron Activation Analysis for Determination of Copper Content of Lower Silesian Copper Deposits. Talanta 11, 1135 (1964).

75) *Greenland, L. P.:* Application of Coincidence Counting to Neutron Activation Analysis. Geological Survey Prof. Paper 600-B, pp. B 76—78. Washington, D. C.: U. S. Geological Survey 1968.

76) *Wyttenbach, A., H. R. Von-Gunten*, and *W. Scherle:* Determination of Bromine Content and Isotopic Composition of Bromine in Stony Meteorites by Neutron Activation. Geochim. Cosmochim. Acta 29, 467 (1965).

⁷⁷) *Herr, W.*, and *R. Wolfle:* A Gamma-Gamma Coincidence Arrangement for Activation-Analysis Determination of Trace Amounts of Selenium and Iridium in Minerals, Nickel, Iron and Platinum Metals. Z. Anal. Chem. *209*, 213 (1965).

⁷⁸) *Fujii, I., A. Tani, H. Muto, K. Ogawa*, and *M. Sato:* A Rapid Method for Praseodymium by 14 MeV Neutron Activation. Application of Gamma-Gamma Coincidence Method to the Rapid Determination of Praseodymium. J. At. Energy Soc. Japan *5*, No. 3, 218 (1963).

⁷⁹) *Hall, T. A.:* Chemical Element Analysis of Radioactive Mixtures in Biological Materials. Nucleonics *12*, No. 3, 34 (1954).

⁸⁰) *Borg, D. C., R. E. Segel, P. Kienle*, and *L. Campbell:* Selective Radioactivation and Multiple Coincidence Spectrometry in the Determination of Trace Elements in Biological Material, Measurement of Manganese. Intern. J. Appl. Radiation Isotopes *11*, 10 (1961).

⁸¹) *Nielson, J. M.*, and *H. A. Kornberg:* Multidimensional Gamma-Ray Spectrometry and its Use in Biology. Radioisotope Sample Measurement Techniques in Medicine and Biology, pp. 3—15. Vienna: International Atomic Energy Agency 1965.

⁸²) *Wainerdi, R. E.*, and *M. P. Menon:* Comparison of Nuclear Activation Methods for Bromine. Instrumental Analysis of Pesticides in Plant Materials Close to the Limit of Sensitivity. Proceedings of the 1967 Symposium, Nuclear Activation Techniques in the Life Sciences, Amsterdam, May 8—12, 1967, pp. 33—50. Vienna: International Atomic Energy Agency 1967.

⁸³) *Wahlgren, M. A.:* Application of Special Counting Techniques to Forensic Problems. General Atomics Report 8171, pp. 79—102 (1967).

⁸⁴) *Coleman, R. F.:* The Determination of Lithium by Neutron Activation. Analyst *85*, 285 (1960).

⁸⁵) *Brune, D., S. Mattsson*, and *K. Liden:* Application of a Betatron in Photonuclear Activation Analysis. The 1968 International Conference Modern Trends in Activation Analysis, Gaithersburg, Maryland, October 7—11, 1968, Paper 2.

⁸⁶) *Albert, P.:* The Use of Reactions Induced by Accelerated Protons, Deuterons, Helions and Gamma Photons in Radioactivation Analysis for the Determination of Oxygen, Carbon, and Nitrogen in Metals. Modern Trends in Activation Analysis-Proceedings of the 1961 International Conference in College Station, Texas, December, 1961, A and M College of Texas, College Station, Texas, pp. 78—85 (1961).

⁸⁷) — Radiochemical Methods in Analytical Chemistry. Activation Analysis with Gamma-Ray Photons and Charged Particles. Chimica *21*, No. 3, 116 (1967).

⁸⁸) *Engelmann, C., J. Gosset, M. Loeuillet, A. Marschal, P. Ossart*, and *M. Boissier:* Examples of Determination of Light Elements in Various High Purity Materials by Gamma Photon and Charged Particle Activation. The 1968 International Conference Modern Trends in Activation Analysis, Gaithersburg, Maryland, October 7—11, 1968, Paper 58.

⁸⁹) *Barrandon, J. N.*, and *P. Albert:* Determination of Oxygen Present at the Surface of Metals by Irradiation with 2 MeV Tritons. The 1968 International Conference Modern Trends in Activation Analysis, Gaithersburg, Maryland, October 7—11, 1968, Paper 93.

⁹⁰) *Tilbury, R. S.*, and *W. H. Wahl:* Activation Analysis by High-Energy Particles. Nucleonics *23*, No. 9, 70 (1965).

⁹¹) *Debrun, J. L., J. N. Barrandon*, and *P. Albert:* Contribution to Activation Analysis by Charged Particles; Determination of Carbon and Oxygen in Pure Metals, Possibilities of Sulphur Determination. The 1968 International Con-

ference Modern Trends in Activation Analysis, Gaithersburg, Maryland, October 7—11, 1968, Paper 95.

92) *Nazaki, T., N. Yatsurugi, N. Akiyana,* and *I. Imai:* Charged Particle Activation Analysis for Carbon, Nitrogen and Oxygen in Semiconductor Silicon. The 1968 International Conference Modern Trends in Activation Analysis, Gaithersburg, Maryland, October 7—11, 1968, Paper 8.

93) *Greenwood, R. C.,* and *J. Reed:* Scintillation Spectrometer Measurements of Capture Gamma-Rays from Natural Elements. Modern Trends in Activation Analysis—Proceedings of the 1961 International Conference in College Station, Texas, December, 1961. A and M College of Texas, College Station, Texas, pp. 166—171 (1961).

94) *Lombard, S. M., T. L. Isenhour, P. H. Heintz, G. L. Woodruff,* and *W. E. Wilson:* Neutron Capture Gamma-Ray Activation Analysis. Intern. J. Appl. Radiation Isotopes *19,* 15 (1968).

95) *Lussie, W. G.,* and *J. L. Brownlee:* The Measurement and Utilization of Neutron-Capture Gamma Radiation. Modern Trends in Activation Analysis—Proceedings of the 1965 International Conference in College Station, Texas, April, 1965. A and M College of Texas, College Station, Texas, pp. 194—199 (1965).

96) *Pierce, T. B., P. F. Peck,* and *W. M. Henry:* Determination of Carbon in Steels by Measurement of the Prompt Gamma-Radiation Emitted During Proton Bombardment. Nature *204,* 571 (1964).

97) — — — The Rapid Determination of Carbon in Steels by Measurement of the Prompt Radiation Emitted During Deuteron Bombardment. Analyst *90,* 339 (1965).

98) *Reinig, W. C.,* and *A. G. Evans:* Californium-252: A New Neutron Source for Activation Analysis. The 1968 International Conference Modern Trends in Activation Analysis, Gaithersburg, Maryland, October 7—11, 1968, Paper 33.

99) *Lutz, G. J.* Calculation of Sensitivities in Photon Activation Analysis. Anal. Chem. *41,* 424 (1969).

100) *Seagrave, J. D.:* U.S. Atomic Energy Commision Report, LAMS-2162, 1958.

101) *Brownlee, J. L.:* The Detection and Determination of Fissionable Species by Neutron Activation — Delayed Neutron Counting. The 1968 International Conference Modern Trends in Activation Analysis, Gaithersburg, Maryland, October 7—11, 1968, Paper 166.

102) *Gale, N. H.:* Development of Delayed Neutron Technique as a Rapid and Precise Method for Determination of Uranium and Thorium at Trace Levels in Rocks and Minerals, with Applications to Isotope Geochronology. Radioactive Dating and Methods of Low Level Counting, pp. 431—452. Vienna: International Atomic Energy Agency 1967.

103) *Amiel, S., J. Gilat,* and *D. Heymann:* Uranium Content of Chondrites by Thermal Neutron Activation and Delayed Neutron Counting. Geochim. Cosmochim. Acta *31,* 1499 (1967).

Received May 27, 1969

The Fragmentation of Transition Metal Organometallic Compounds in the Mass Spectrometer

R. B. King

Research Professor, Department of Chemistry, University of Georgia,
Athens, Georgia 30601, U.S.A.

Contents

A. Introduction 92
B. Some General Features of the Mass Spectra of Transition Metal Organo-
   metallic Compounds 93
C. Mass Spectra of Polynuclear Metal Carbonyl Derivatives 95
D. Mass Spectra of Cyclopentadienylmetal Carbonyl Derivatives 96
E. Mass Spectra of Olefin Metal Carbonyl Derivatives 103
F. Mass Spectra of Tertiary Phosphine Derivatives of Metal Carbonyls 105
G. Mass Spectra of Metal Carbonyl Halides and Metal Carbonyl Hydrides .. 107
H. Mass Spectra of Alkylthio, Dialkylphosphido, and Dialkylureylene Deriva-
   tives 110
I. Mass Spectra of Fluorocarbon Derivatives of Transition Metals 113
J. Mass Spectra of π-Cyclopentadienyl Derivatives without Carbonyl Ligands 116
K. References 124

A. Introduction

Since the discovery of ferrocene in 1951 transition metal organometallic chemistry [1] has received much attention in numerous laboratories throughout the world. Compounds with a variety of unusual structures and properties have been prepared. Some of these compounds are of practical interest as catalysts for the synthesis of unusual and useful organic compounds [2].

As the development of transition metal organometallic chemistry has progressed resulting in the knowledge of an increasing variety of unusual compound types, the emphasis in this field has shifted from the synthesis of new compounds to the study of known compounds by physical and spectroscopic techniques. Mass spectroscopy first began to be applied extensively to the study of transition metal organometallic compounds in 1965, although scattered papers in this area had appeared as early as 1955 including particularly *Friedman*, *Irsa*, and *Wilkinson*'s [3] now classic study on the mass spectroscopy of biscyclopentadienylmetal derivatives. Recently several comprehensive reviews have appeared in the area of mass spectroscopy of organometallic compounds [4,5].

This present article summarizes the results of these studies as well as related mass spectroscopic data obtained in the author's laboratories as well as some related studies by other workers [4,5]. For a review of the general principles of mass spectrometry the reader is referred to any of several books in this area [6-10] (see also [11-23] on information to be obtained from mass spectra).

B. Some General Features of the Mass Spectra of Transition Metal Organometallic Compounds

The majority of transition metal organometallic compounds contain carbonyl and/or cyclopentadienyl groups. Before reviewing the fragmentation patterns of selected transition metal organometallic compounds with more unusual ligands, some characteristics of the fragmentation patterns of metal carbonyls and cyclopentadienyls will first be examined. This will enable the remaining discussion to focus on the more unusual aspects of the fragmentation patterns of compounds with other structural features.

The mass spectra of simple *metal carbonyls* such as $M(CO)_6$, ($M = Cr$, Mo, and W) [13], $Fe(CO)_5$ [12], and $Ni(CO)_4$ [12] exhibit stepwise loss of carbonyl groups from the molecular ion to give ions of the type $M(CO)_n^+$. This metal-carbon bond cleavage is such a favored fragmentation pathway that it predominates over most other alternatives in the mass spectra of metal carbonyls containing other ligands. Thus, in most cases the molecular ion loses all of its carbonyl groups before any types of fragmentation processes take place.

An alternate fragmentation pathway for metal carbonyl derivatives involves cleavage of the carbon-oxygen bonds rather than the metal-carbon bonds. This type of process is only observed in metal carbonyls with unusually strong metal-carbon bonds. As carbonyl groups are successively lost, the remaining metal-carbon bonds become stronger[a] since the available electron density for retrodative bonds is shared between fewer carbonyl groups. For this reason, carbon-oxygen bond

---

[a] If the metal-carbon bonds did not become stronger upon successive loss of carbonyl groups, stepwise loss of carbonyl groups would not be observed. Instead all carbonyl groups in a metal carbonyl would be lost almost simultaneously in contrast to all reported observations. There is some evidence for the simultaneous loss of two carbonyl groups in the mass spectra of some metal carbonyl derivatives with particularly strong metal-carbon bonds suggesting that as the metal-carbon bond becomes stronger the difference in energy required to remove sucessive carbonyl groups descreases. This observation suggests that the effect of additional retrodative bonding is less in cases where there is already a relatively large amount of retrodative bonding.

cleavage can compete with metal-carbon bond cleavage in the mass spectra of metal carbonyl derivatives only after most of the carbonyl groups are already lost. The ions WC^+ and WC_2O^+ arising from carbon-oxygen bond cleavage are found in the mass spectrum of hexacarbonyltungsten, but this type of carbon-oxygen bond cleavage cannot compete effectively with metal-carbon bond cleavage in ions containing more carbonyl groups.

The *metal-π-cyclopentadienyl bond* is somewhat stronger in π-cyclopentadienyl derivatives than the metal-carbon bond in metal carbonyl derivatives. However, stepwise loss of π-cyclopentadienyl ligands occurs in the mass spectra of π-cyclopentadienyl derivatives [24]. The following fragmentation of ferrocene is particularly important because of the occurrence of ferrocene as a pyrolysis product in numerous mass spectra:

$$(C_5H_5)_2Fe^+ \xrightarrow[\text{m* 78.5}]{-C_5H_5} C_5H_5Fe^+ \xrightarrow[\text{m* 25.8}]{-C_5H_5} Fe^+$$
$$\text{m/e 186} \qquad\qquad \text{m/e 121} \qquad\qquad \text{m/e 56}$$

Elimination of a neutral C_2H_2 fragment from the C_5H_5 ligand also occurs frequently in the mass spectra of π-cyclopentadienyl derivatives especially in processes of the following type [24]:

$$C_5H_5M^+ \xrightarrow{-C_2H_2} C_3H_3M^+ \xrightarrow{-C_3H_3} M^+$$

The molecular ion of cobaltocene also undergoes elimination of a methyl group by the following sequence [24]:

$$(C_5H_5)_2Co^+ \rightarrow C_9H_7Co^+ + CH_3$$

This process must clearly involve a hydrogen shift.

In one of the earliest detailed mass spectroscopic studies with transition metal organometallic compounds, *Friedman, Irsa,* and *Wilkinson* [3] showed that in the mass spectra of $(C_5H_5)_2M$ derivatives the molecular ion was more stable in the covalent π-cyclopentadienyls than in the ionic cyclopentadienides.

The mass spectra of some transition metal organometallic derivatives sometimes exhibit dipositive ions which are identified by peaks at half-integral m/e values. Dipositive ions are much more intense in the mass spectra of organometallic derivatives of the third row ($5d$) transition metals (e. g. tungsten) than in the mass spectra of the corresponding organometallic derivatives of the lighter ($3d$ and $4d$) transition metals. Tripositive ions are extremely rare, but there is some evidence for the occurrence of the ion $C_5H_5W(CO)_2^{+++}$ [19].

C. Mass Spectra of Polynuclear Metal Carbonyl Derivatives

The facile loss of carbonyl groups in the mass spectra of metal carbonyls permits the generation of novel bare metal cluster ions M_x^+ in the mass spectra of polynuclear metal carbonyls of the type $M_x(CO)_y$. Thus in the mass spectra of $Mn_2(CO)_{10}$ and $Co_2(CO)_8$ all carbonyl groups are lost before rupture of the metal-metal bond resulting in the production of the bimetallic ions Mn_2^+ and Co_2^+, respectively [14].

The trimetallic ions M_3^+ ($M = Ru$ and Os) are major components in the mass spectra of $M_3(CO)_{12}$ ($M = Ru$ and Os) [25,26]. The tetrametallic ions M_4^+ ($M = Co$ and Rh) similarly are major components in the mass spectra of $M_4(CO)_{12}$ ($M = Co$ or Rh) [25,26]. The Os_4^+ cluster is found in the mass spectrum of $Os_4O_4(CO)_{12}$ [25]. Further fragmentation of these M_x^+ metal clusters results in metal-metal bond rupture giving the ions M_n^+ ($1 \leq n \leq x$).

In some polynuclear metal carbonyls of the first row ($3d$) transition metals the metal-metal bonds are too weak to survive complete loss of carbonyl groups. Thus in the mass spectrum of $Fe_3(CO)_{12}$ stepwise loss of carbonyl groups occurs only as far as the tricarbonyltriiron ion $Fe_3(CO)_3^+$ [26]. The rupture of the iron-iron bonds competes with the stepwise loss of carbonyl groups giving ions such as $Fe_2(CO)_4^+$ and $Fe(CO)_4^+$. The mass spectrum of $Fe_2(CO)_9$ exhibits the ion $Fe(CO)_5^+$ which may represent pentacarbonyliron formed by the following pyrolysis process [25]:

$$3 \ Fe_2(CO)_9 \rightarrow 3 \ Fe(CO)_5 + Fe_3(CO)_{12}$$

Mass spectral data suggest that metal-metal bonds between first row transition metals are weaker than corresponding metal-metal bonds between second and third row transition metals. Thus in the series

$$(CO)_5 M - M'(CO)_5 \ (M = M' = Mn \ or \ Re; \ M = Mn, \ M' = Re)$$

the metal-metal bond strengths increase in the following sequence [27]

$$Mn-Mn \ (least) < Re-Re < Mn-Re \ (greatest).$$

The greater strength of the bond between dissimilar metal atoms may arise from an electronegativity difference between the dissimilar metal atoms which adds an electrostatic force to strengthen the metal-metal bond.

The mass spectra of some compounds containing metal-carbon clusters have been investigated. The mass spectra of the $YCCo_3(CO)_9$ compounds (7: $Y = H$, F, Cl, Br, CH_3 and C_6H_5) exhibit stepwise loss of nine CO groups giving the interesting cluster ions $YCCo_3^+$ [26,28,29]. The ion $CH_3CCo_3^+$ from $CH_3CCo_3(CO)_9$ (7: $Y = CH_3$) undergoes further dehydrogenation to give $HC_2Co_3^+$, metal-carbon bond cleavage to give HCo_3^+ and Co_3^+, and cobalt-cobalt bond cleavage in these latter fragments to give ions with one or two cobalt atoms. The phenyl derivative

$C_6H_5CCo_3(CO)_9$ (1: $Y = C_6H_5$) [28)]

exhibits the series of doubly charged ions $C_6H_5CCo_3(CO)_{6-n}^{2+}$. The mass spectrum of $[CCo_3(CO)_9]_2$ (2) exhibits stepwise loss of all eighteen CO groups to give the unusual metal carbide ion $C_2Co_6^+$ [29)].

Another source of metal carbide ions is the ruthenium compound $Ru_6C(CO)_{17}$. The mass spectrum of this compound exhibits an ion series $Ru_6C(CO)_n^+$ ($n = 0$ thru 16)[30)]. A low abundance of Ru_5C^+ is also found[30)]. The mass spectra of the $Ru_6(CO)_{14}$ (arene) derivatives have also been investigated. A series of ions $Ru_6C(CO)_{14-n}$ (arene)$^+$ is observed [30)].

D. Mass Spectra of Cyclopentadienylmetal Carbonyl Derivatives

A variety of interesting effects have been noted in the mass spectra of cyclopentadienylmetal carbonyl derivatives [19,31)]. Cyclopentadienyliron carbonyl derivatives have a great tendency to undergo pyrolysis in the mass spectrometer producing ferrocene which gives rise to an ion $(C_5H_5)_2$-Fe^+ (m/e 186). Furthermore, compounds of the type $RFe(CO)_2C_5H_5$ produce substituted ferrocenes of the type $C_{10}H_9RFe$ and $C_{10}H_8R_2Fe$ by similar pyrolytic processes. Certain acyl derivatives

(e.g. $C_6H_5COFe(CO)_2C_5H_5$)

undergo decarbonylation in the mass spectrometer to produce the corresponding alkyl derivatives; such decarbonylation reactions have been previously effected photochemically.

The ions $C_5H_5Fe(CO)_nI^+$ ($n = 2$, 1, and 0) have been observed in the mass spectra of the

$RFe(CO)_2C_5H_5$ derivatives ($R = C_6H_5COS$, CH_3S, $NC_5H_4CH_2$, $C_5H_{10}NCH_2CH_2$, CH_3OCOCH_2, C_6H_5, and $C_6H_5CH=CHCO$).

The ions $C_5H_5Mo(CO)_nI^+$ ($n = 3, 2, 1$, and 0) have been observed in the mass spectra of the

$RMo(CO)_3C_5H_5$ derivatives ($R = CH_3SCH_2$, CH_2NCO, $1/2 (CF_2)_3$, and $Br(CH_2)_4$).

The mass spectrum of $[C_5H_5Cr(NO)_2]_2$ exhibits the ions $C_5H_5Cr(NO)_nI^+$ ($n = 2, 1$, and 0).

Some cyclopentadienyliron carbonyl derivatives exhibit sufficient metastable ions for the complete fragmentation scheme from the molecular ion down to the bare iron ion Fe^+ to be elucidated [19]. One such compound is the acetyl derivative $CH_3COFe(CO)_2C_5H_5$ *(3)* (Fig. 1). The parent ion $CH_3COFe(CO)_2C_5H_5^+$ of this compound appears to break down via two major pathways. In the first pathway successive losses of carbonyl groups occur to give the carbonyl-free ion $CH_3FeC_5H_5$. This ion then loses hydrogen to give the ion $C_6H_6Fe^+$ which subsequently loses a C_6H_6 fragment to give the bare ion Fe^+. In the second pathway for the fragmentation of the molecular ion $CH_3COFe(CO)_2C_5H_5^+$ loss of a methyl group first occurs giving the ion $C_5H_5Fe(CO)_3^+$. This ion then loses carbonyl groups stepwise giving the ion $C_5H_5Fe^+$ which finally loses the cyclopentadienyl ring to give again the bare iron ion Fe^+.

Another cyclopentadienyliron carbonyl derivative for which complete fragmentation schemes could be deduced by metastable ion analysis is the methyl ester $CH_3OCOCH_2Fe(CO)_2C_5H_5$ *(4)* (Fig. 2) [19]. Again the parent ion $CH_3OCOCH_2Fe(CO)_2C_5H_5^+$ appears to break down by two major pathways. In the first pathway losses of carbonyl groups occur to give the ion $CH_3OCOCH_2FeC_5H_5^+$. Analysis of the metastable ions shows that the two carbonyl groups can be lost from the molecular ion either individually or simultaneously. Further fragmentation of the $CH_3OC-OCH_2FeC_5H_5^+$ ion occurs by the elimination of ketene and concurrent shift of a methoxy group from carbon to iron to give the ion $C_5H_5FeOCH_3^+$.

This ion then undergoes dehydrogenation to form the ion

$C_5H_5FeCOH^+$

which is then decarbonylated to give $C_5H_6Fe^+$. Loss of cyclopentadiene from the latter ion converts it to the bare iron ion Fe^+. The second pathway for the degradation of the molecular ion

$CH_3OCOCH_2Fe(CO)_2C_5H_5^+$

3   4   5   6

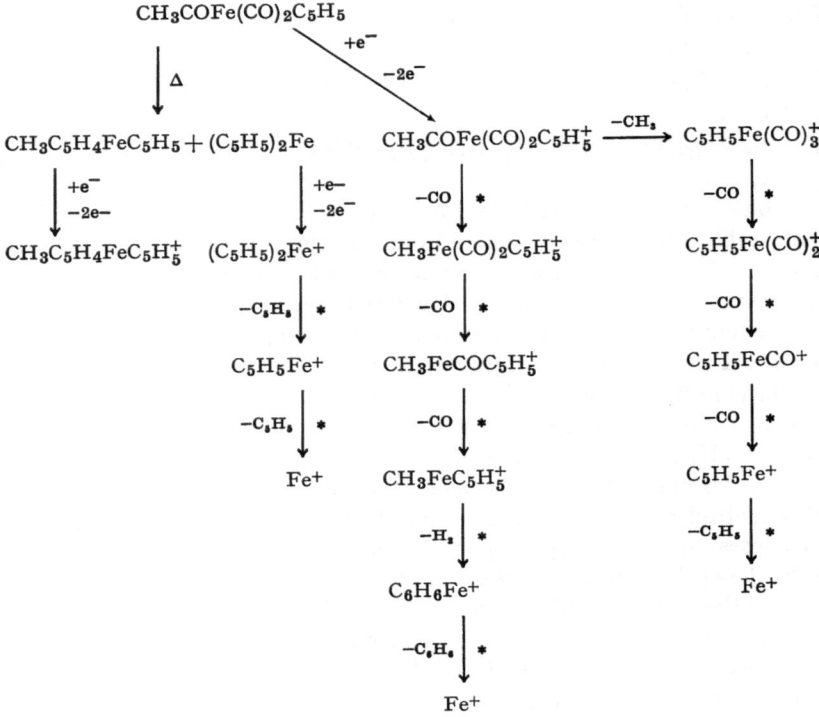

* The metastable ion corresponding to this process was observed.

Fig. 1. Fragmentation scheme of $CH_3COFe(CO)_2C_5H_5$

first involves loss of the methoxy group to give the ion

$C_5H_5Fe(CO)_2CH_2CO^+$,

possibly of structure *(5)* containing π-bonded ketene. This ion subsequently loses ketene to form the $C_5H_5Fe(CO)_2^+$ ion which is then degraded to Fe^+ by the expected successive losses of the two carbonyl groups followed by loss of the π-cyclopentadienyl ring. Similar ketene-elimination processes are noted in the mass spectrum of the molybdenum derivative $C_2H_5OCOCH_2Mo(CO)_3C_5H_5$ *(6)*.

The mass spectrum of the π-allyl derivative $C_3H_5Mo(CO)_2C_5H_5$ *(7)* also exhibits some interesting features. The molecular ion first loses one of its carbonyl groups to give the monocarbonyl ion $C_3H_5MoCOC_5H_5^+$ which still appears to contain the π-allyl ligand. Fragmentation of this monocarbonyl ion next involves a loss of 30 mass units rather than the usual 28 mass units corresponding to loss of a carbonyl group. This

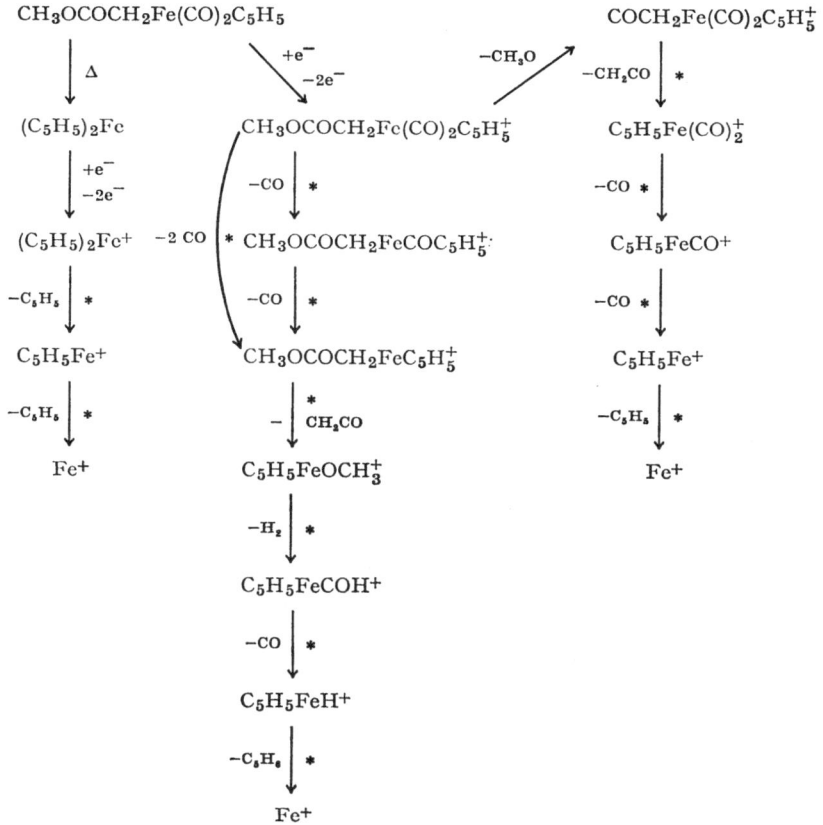

* The metastable ion corresponding to this process was observed.

Fig. 2. Fragmentation scheme of $CH_3OCOCH_2Fe(CO)_2C_5H_5$

indicates that the ion $C_3H_5MoCOC_5H_5^+$ undergoes nearly simultaneous loss of carbon monoxide and hydrogen to give an ion $C_3H_3MoC_5H_5^+$ apparently containing a π-cyclopropenyl group. A similar fragmentation scheme is also noted for the tungsten analogue $C_3H_5W(CO)_2C_5H_5$ (7: M = W) but not for the iron compound $C_3H_5FeCOC_5H_5$ [32].

The mass spectra of binuclear cyclopentadienylmetal carbonyl derivatives of chromium and molybdenum provide useful indications of the relative strengths of the metal-metal bonds in various compound types. The mass spectrum of $[C_5H_5Cr(CO)_3]_2$ (8: M = Cr) exhibits no bimetallic ions but only monometallic ions such as $C_5H_5Cr(CO)_n^+$ ($n = 3$,

2, 1, and 0) as well as the ions $C_5H_5Cr(CO)_nH^+$ ($n = 3$, 2, and 1) formed by proton abstraction from the mass spectrometer [26]. This indicates that the metal-metal bond in $[C_5H_5Cr(CO)_3]_2$ is so weak that it breaks upon vaporization in the mass spectrometer. The mass spectrum of the molybdenum compound $[C_5H_5Mo(CO)_3]_2$ (8: M = Mo) exhibits the bi-

7   8

metallic ions $(C_5H_5)_2Mo_2(CO)_n^+$ ($n = 6$, 5, 4, 3, 2, and 0) as well as some monometallic ions indicating that the molybdenum-molybdenum bond is strong enough to survive vaporization and remains partially intact upon electron impact [26,31]. The mass spectra of $[(CH_3)_5C_5Mo(CO)_2]_2$ (9) and $[C_9H_9Mo(CO)_2]_2$ (10) both show only bimetallic ions indicating a very strong metal-metal bond consistent with the presence of metal-metal triple bonds as indicated in the structures 9 and 10 [34].

The mass spectra of $[C_5H_5M(CO)_2]_2$ (M = Fe [31] and Ru [33]) and of $[C_5H_5Cr(NO)_2]_2$ [33] have been investigated. The carbonyl derivatives exhibit only rather routine decarbonylation, dehydrogenation, and acetylene-elimination processes. The nitrosyl derivative $[C_5H_5Cr(NO)_2]_2$ exhibited stepwise loss of nitrosyl groups [33].

A particularly interesting cyclopentadienylmetal carbonyl derivative is the iron compound $[C_5H_5FeCO]_4$ which has an unusual symmetrical tetrahedral structure and a $\nu(CO)$ frequency at ~ 1620 cm^{-1} indicative of an unusually strong iron-carbon bond and an unusually weak carbon-oxygen bond [35]. The mass spectrum of $[C_5H_5FeCO]_4$ exhibits a molecular ion but very weak carbonyl-containing tetrametallic fragment ions $(C_5H_5)_4Fe_4(CO)_n^+$ ($n = 3$, 2, and 1) and $(C_5H_5)_4Fe_4(CO)_nC^+$ ($n = 2$ and 0); the relative intensities of these two series of ions are approximately equal indicating similar tendencies for iron-carbon and carbon-oxygen cleavage in contrast to most metal carbonyl derivatives where metal-carbon cleavage occurs much more easily than carbon-oxygen cleavage. This anomalous behaviour in the mass spectrum of $[C_5H_5FeCO]_4$ may clearly be attributed to the unusually weak carbon-oxygen bonds and the unusually strong iron-carbon bonds in this complex. Trimetallic ions observed in the mass spectrum of $[C_5H_5FeCO]_4$ include $(C_5H_5)_3Fe_3C^+$ and the series $(C_5H_5)_3Fe_3(CO)_n^+$ ($n = 2$, 1, and 0); the highest m/e tri-metallic ion $(C_5H_5)_3Fe_3(CO)_2^+$ probably arises by elimination of a neutral $C_5H_5Fe(CO)_2$ fragment from the parent ion $(C_5H_5)_4Fe_4(CO)_4^+$. The tri-

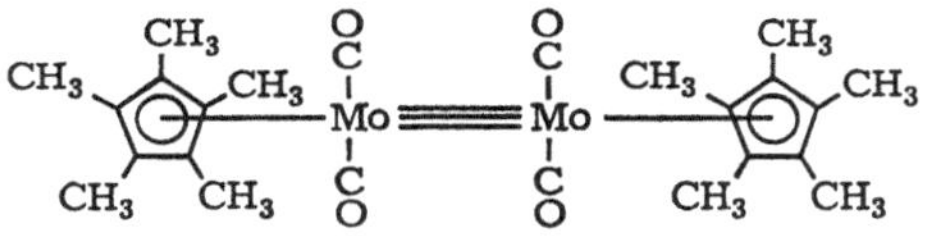

9

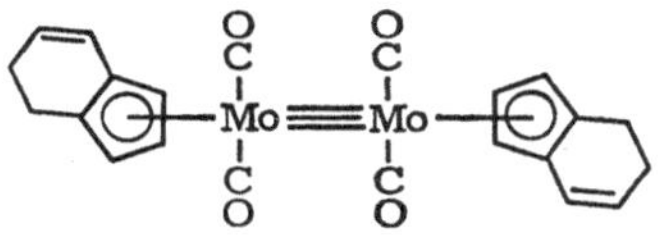

10

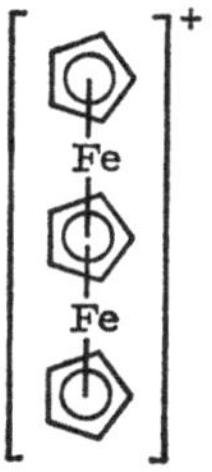

11

metallic ion $(C_5H_5)_3Fe_3^+$ undergoes further fragmentation by the following iron-elimination process [36]:

$$(C_5H_5)_3Fe_3^+ \xrightarrow[m^* 259.8]{} (C_5H_5)_3Fe_2^+ + Fe$$
$$m/e\ 363 \qquad\qquad m/e\ 307$$

The $(C_5H_5)_3Fe_2^+$ ion may have a *triple decker sandwich structure* such as *11*. This possibility is supported by its further fragmentation by elimination of a neutral C_5H_5Fe fragment by the following process:

$$(C_5H_5)_3Fe_2^+ \xrightarrow[m^* 112.4]{} (C_5H_5)_2Fe^+ + C_5H_5Fe$$
$$me/\ 307 \qquad\qquad m/e\ 186$$

This is the only example of an elimination of a neutral C_5H_5M fragment supported by the presence of a metastable ion which has been observed in work with a great variety of π-cyclopentadienyl derivatives.

Similar triple-decker sandwich ions $(C_5H_5)_3M_2^+$ (M = Fe or Ni) have been observed in very low abundances in the mass spectra of ferrocene and nickelocene at pressures high enough for bimolecular reactions to occur [23].

Müller and *Herberhold* [37] have studied the mass spectra of some cyclopentadienylmanganese carbonyl derivatives of the type

$C_5H_5Mn(CO)_2L$

(L = CO, cycloolefins, maleic anhydride, isocyanides, amines, phosphines, sulfoxides, etc.). In most cases loss of the two carbonyl groups occurs in a single step. The ion $C_5H_5MnC_5H_8^+$ from the cyclopentene derivative undergoes dehydrogenation to give the ion

$(C_5H_5)_2Mn^+$.

The ion $C_5H_5MnCNC_6H_{11}^+$ from the cyclohexyl isocyanide complex undergoes successive losses of hydrogen cyanide and cyclohexene in either possible sequence. The following processes exemplify unusual fragmentations observed in the mass spectrum of the dimethylsulfoxide complex:

$$C_5H_5MnSO(CH_3)_2^+ \rightarrow C_5H_5MnO^+ + (CH_3)_2S$$
$$C_5H_5MnSOCH_3^+ \rightarrow C_5H_5MnOH^+ + S{=}CH_2$$

The mass spectra of some π-indenylmolybdenum carbonyl derivatives of the type $RMo(CO)_nC_9H_7$ have been investigated [38]. The usual stepwise loss of carbonyl groups is observed except for the π-allyl derivative $\pi\text{-}C_3H_5Mo(CO)_2C_9H_7$ where the loss of H_2 accompanies the loss of the last carbonyl group as in the related π-allyl derivative

$\pi\text{-}C_3H_5Mo(CO)_2C_5H_5$ (7).

In the mass spectrum of the σ-methyl derivative

$CH_3Mo(CO)_3C_9H_7$

loss of the methyl group competes with loss of the last carbonyl group to give the ion $C_{10}H_8Mo^+$. The ion $C_9H_7Mo^+$ is a major fragment in the mass spectra of some π-indenylmolybdenum carbonyl derivatives; it apparently can undergo two successive C_2H_2 eliminations. The metal-free indenyl ion $C_9H_7^+$ also undergoes easily two successive C_2H_2 eliminations. The mass spectrum of the π-allyl derivative

$\pi\text{-}C_3H_5Mo(CO)_2C_9H_7$

indicates the presence of about 2% of a previously unidentified impurity of empirical formula corresponding to

$\pi\text{-}C_3H_5Mo(CO)_2(\pi\text{-}C_9H_6C_3H_5)$;

this species exhibits a fragmentation pattern similar to that of the other π-allylmolybdenum carbonyl derivatives including the loss of H_2 with the loss of the last carbonyl group.

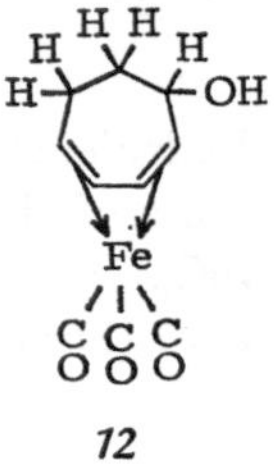

12

E. Mass Spectra of Olefin Metal Carbonyl Derivatives

The mass spectra of monometallic olefin derivatives of metal carbonyls of the following types have been investigated:

a) (triene)$M(CO)_3$: $M = Cr$, triene = benzene [39,40], cycloheptatriene [40], cyclooctatetraene [40]; $M = W$, triene = toluene [40], p-xylene [40], mesitylene [40], cycloheptatriene [40] 1,3,5-cyclooctatriene [40];

b) (diene)$M(CO)_4$: diene = 1,5-cyclooctadiene, $M = Mo$ and W [40]; bicyclo[2,2,1]heptadiene, $M = W$ [40]; dicyclopentadiene, $M = W$ [40];

c) (diene)$_2M(CO)_2$: diene = 1,3-cyclohexadiene, $M = Mo$ or W [40];

d) (diene)$Fe(CO)_3$: diene = cyclooctatetraene [40], 1,3-cyclohexadiene [41], and cycloheptadienol [40].

Stepwise loss of carbonyl groups are observed in these mass spectra. The carbonyl-free ions then generally undergo further fragmentations by dehydrogenation and elimination of two carbon fragments, particularly C_2H_2. In cases where the olefin has two adjacent sp^3 carbon atoms (1,3-cyclohexadiene, 1,3,5-cyclooctatriene, and 1,5-cyclooctadiene) dehydrogenation is particularly facile and occurs prior to the loss of all carbonyl groups. The mass spectrum of the cycloheptadienol derivative $C_7H_9OHFe(CO)_3$ *(12)* exhibits facile elimination of H_2O.

The diolefins bicyclo[2.2.1] heptadiene and dicyclopentadiene are Diels-Alder adducts of cyclopentadiene. In the mass spectra of the tungsten carbonyl complexes of these olefins a retro-Diels-Alder fragmentation (C_5H_6 elimination) is observed to take place with such facility that it is observed before all carbonyl groups are lost. Thus the mass spectrum of bicyclo[2,2,1]-heptadienetetracarbonyltungsten *(13)* exhibits the family of ions $C_2H_2W(CO)_n^+$ ($n = 2$, 1, and 0) formed by bond cleavage as indicated by dotted lines in structure *13*. Similarly, the mass spectrum of dicyclopentadienetetracarbonyltungsten *(14)* exhibits the family of ions $C_5H_6W(CO)_n^+$ ($n = 0$, 1, 2, or 3) formed by bond cleavage as indicated by the dotted lines on structure *14*.

13   14

The mass spectra of some binuclear iron carbonyl complexes of the type (triene) $Fe_2(CO)_6$ have been investigated [26,33]. Reaction between

$Fe_3(CO)_{12}$, 1,4-dibromobutyne-2, and zinc gives a volatile air-stable red-orange solid, m. p. 69—70 °, originally [42] believed to be $C_4H_4Fe_2(CO)_5$ but later [26] shown by mass spectrometry to be $C_4H_4Fe_2(CO)_6$; this compound appears to have structure *15*. The molecular ion of this complex undergoes stepwise loss of all six carbonyl groups. The resulting $C_4H_4Fe_2^+$ ion then loses its C_4H_4 group giving the bare diiron ion Fe_2^+ which fragments further giving an iron atom and a bare iron ion Fe^+. A complete series of metastable ions is observed for all eight of these steps.

The mass spectrum of the 1,3,5-cyclooctatriene complex

$C_8H_{10}Fe_2(CO)_6$

has also been investigated [33]. The molecular ion in the mass spectrum of $C_8H_{10}Fe_2(CO)_6$ undergoes the stepwise loss of its six carbonyl groups giving the ions $C_8H_{10}Fe_2(CO)_n^+$ ($n = 5, 4, 3, 2, 1$, and 0). In some cases one iron atom can also be lost forming the ions $C_8H_{10}Fe(CO)_n^+$ ($n = 1$ and 0 and possibly 3 and 2). The end product from the loss of all six carbonyl groups and one iron atom from the molecular ion is the carbonyl-free monometallic ion $C_8H_{10}Fe^+$. Metastable ion analysis indicates that the ion $C_8H_{10}Fe^+$ goes to the bare iron ion Fe^+ by the following two steps:

$$C_8H_{10}Fe^+ \rightarrow C_6H_6Fe^+ + C_2H_4$$
$$C_6H_6Fe^+ \rightarrow Fe^+ + C_6H_6$$

Other iron-containing ions in the mass spectrum of $C_8H_{10}Fe_2(CO)_6$ arising from less important fragmentation processes include $C_5H_5Fe^+$, Fe_2^+, and $FeCO^+$.

The mass spectra of binuclear iron carbonyl complexes of acenaphthylene and azulene have been briefly reported [26]. The acenaphthylene complex previously [43] reported as $C_{12}H_8Fe_2(CO)_6$ exhibits $C_{12}H_8Fe_2(CO)_5^+$ as the highest m/e ion; the implied pentacarbonyl formulation was later [44] confirmed by X-ray crystallography which indicated structure *16*. The reaction between azulene and $Fe(CO)_5$ has been reported [45] to give dark red $C_{10}H_8Fe_2(CO)_5$; this formulation was confirmed by

15

16

mass spectroscopy [26]. However, under slightly different conditions azulene and $Fe(CO)_5$ gave another dark red product which was shown by mass spectrometry [26] to be the previously unknown $[C_{10}H_8Fe(CO)_2]_2$. The mass spectra of these olefin-iron carbonyl complexes exhibited not only the corresponding molecular ions but also fragment ions corresponding to the stepwise loss of all carbonyl groups.

F. Mass Spectra of Tertiary Phosphine Derivatives of Metal Carbonyls

The relatively low molecular weight of the tris(dimethylamino)phosphine ligand gives its metal carbonyl complexes [45] appreciable volatility thereby making them suitable for mass spectral studies. The mass spectra of several tris(dimethylamino)phosphine complexes of metal carbonyls have been reported [46,47]. The usual stepwise loss of carbonyl groups occurs in the mass spectra of these compounds though sometimes two carbonyl groups are lost in a single step.

Another process occurring in the fragmentation of tris(dimethyl-amino)phosphine-metal carbonyl complexes is the loss of a $(CH_3)_2N$ group from a $[(CH_3)_2N]_3PM(CO)_n^+$ ion to form the corresponding $[(CH_3)_2N]_2PM(CO)_n^+$ ion. This process competes effectively with the loss of carbon monoxide from the same ion. A rough indication of the relative tendencies for the molecular tris(dimethylamino)phosphine-metal carbonyl ion (P^+) to lose carbon monoxide and to lose a dimethylamino group is provided by the ratio $[P-44]/([P-44] + [P-28])$ where $[P-44]$ and $[P-28]$ correspond to the relative intensities of the ions 44 and 28 mass units less than the molecular ion, respectively. The tendency for the molecular ions to lose carbon monoxide to form the $(P-28)^+$ ion is negatively correlated with the metal-carbon bond strength since this process involves rupture of a metal-carbon bond. However, the loss of a dimethyl-amino group from a molecular ion to form the $(P-44)^+$ ion does not involve rupture of a direct metal-ligand bond but instead rupture of a phosphorus-nitrogen bond. Thus the tendency for the molecular ion to lose a dimethylamino group is much less affected by variations in the metal-carbon bond strength than the tendency for the parent ion to lose a carbonyl group. For this reason higher values of the ratio $[P-44]/([P-44] + [P-28])$ may indicate a higher metal-carbon bond strength at least for closely similar compounds. This ratio is much higher for tungsten carbonyl derivatives than for the completely analogous molybdenum and chromium carbonyl derivatives thereby indicating greater strength of tungsten-carbon bonds than similar molybdenum-carbon bonds and chromium-carbon bonds.

Another process of interest in the mass spectra of tris(dimethyl-amino)phosphine derivatives is the elimination of a CH_3NCH_2 (azapropene) fragment. This occurs in ions containing first-row transition metals but no carbonyl groups, e.g.

$$[(CH_3)_2N]_3PM^+ \rightarrow [(CH_3)_2N]_2PHM^+ + CH_3NCH_2 \text{ (M=Cr or Fe)}$$
$$C_5H_5VP[N(CH_3)_2]_3^+ \rightarrow C_5H_5VPH[N(CH_3)_2]_2^+ + CH_3NCH_2$$

Other neutral fragments eliminated in the fragmentation processes of tris-(dimethylamino) phosphine metal carbonyl complexes appear to contain phosphorus-hydrogen bonds, e.g.

$$[(CH_3)_2N]_2PHFe^+ \rightarrow (CH_3NCH_2)_2Fe^+ + PH_3$$
$$[(CH_3)_2N]_3PFe^+ \rightarrow (CH_3NCH_2)_2Fe^+ + (CH_3)_2NPH_2$$

A characteristic feature of the mass spectra of metal carbonyl complexes containing two tris(dimethylamino)phosphine ligands is the presence of ions of the type $[(CH_3)_2N]_4PM^+$ presumably of the structure $[(CH_3)_2N]_3PMN(CH_3)_2^+$ containing a metal-nitrogen bond. Other ions containing metal-nitrogen bonds are observed including $(CH_3)_2NM^+$ (M = Fe or Cr).

The mass spectra of certain metal carbonyl complexes of triphenylphosphine and 1,2-bis(diphenylphosphino)ethane (Pf-Pf) have been investigated [48]. Besides the usual stepwise loss of carbonyl groups, cleavage of the phosphorus-carbon bond occurs. Thus triphenylphosphine complexes exhibit cleavage of the phenyl-phosphorus bond after all carbonyl groups are lost. The 1,2- bis(diphenylphosphino)ethane complexes (e.g. (Pf-Pf)[W(CO)_5]_2 and (Pf-Pf)M(CO)_4) exhibit elimination of the ethylene bridge between two phosphorus atoms.

Phosphorus trifluoride is a weak σ-donor but strong π-acceptor ligand like carbon monoxide. It therefore forms metal trifluorophosphine complexes similar in properties (but somewhat more stable than) the corresponding metal carbonyls. The mass spectrum of $Ni(PF_3)_4$ [49] exhibits successive loss of the PF_3 ligands analogous to the successive loss of carbonyl groups in the mass spectrum of $Ni(CO)_4$ [12]. Ions of the type $Ni(PF_3)_n^+$ also appear to exhibit the loss of a fluorine atom giving ions of the type $Ni(PF_3)_{n-1}PF_2^+$. In the mass spectra of the cobalt trifluorophosphine derivatives $HCo(CO)_n(PF_3)_{4-n}(n = 0,1,2,3,4)$ a similar fragmentation pattern is observed [50].

G. Mass Spectra of Metal Carbonyl Halides and Metal Carbonyl Hydrides

Halogens and hydrogen can be either terminal or bridging ligands in transition metal carbonyl complexes. When bonded to one metal atom in a terminal position, these ligands are readily lost in the fragmentation in the mass spectrometer. However, when bonded to two metal atoms in a bridging position, these ligands remain bonded to the metal atom(s) until an advanced stage of the fragmentation process. This greater difficulty of cleavage of a bridging ligand as compared with a terminal ligand may be attributed to the need to break two metal-ligand bonds for removal of a bridging ligand but only one metal-ligand bond for removal of a terminal ligand.

This difference in the behavior of bridging and terminal halogens has been observed in the mass spectra of the simple metal carbonyl halides [48,51]. The mass spectra of the metal carbonyl halides $M(CO)_5X$ ($X = Cl$, Br, or I; $M = Mn$ or Re) and $Fe(CO)_4I_2$ exhibit competitive carbonyl and halogen losses giving in the cases of the $M(CO)_5X$ derivatives ions of both the types $M(CO)_nX^+$ and $M(CO)_n^+$. However, in the mass spectra of the binuclear metal carbonyl halides with bridging halogen atoms $[M(CO)_4X]_2$ ($M = Mn$, $X = I$; $M = Re$, $X = I$ or Cl) and $[Rh(CO)_2Cl]_2$ all carbonyl groups are lost to give the ions $M_2X_2^+$ before any of the halogens are lost.

The mass spectra of some metal carbonyl halides with π-cyclopentadienyl, π-cycloheptatrienyl, π-allyl and π-indenyl ligands have been investigated [52,38]. The usual losses of carbonyl groups and C_2H_2 fragments are observed. The mass spectra of the chlorides exhibited a strong iodine memory effect, since ions expected for the corresponding iodides were also observed in their mass spectra. Pyrolysis of the halides

$$C_5H_5Mo(CO)_3X$$

to the new binuclear halides $[C_5H_5Mo(CO)X]_2$ occurs in the mass spectrometer. The halides $[C_5H_5Mo(CO)X]_2$ probably have structure *17* somewhat similar to that of $[Rh(CO)_2Cl]_2$; the central metal atoms in the two types of complexes are isoelectronic. Because of a combination of pyrolysis and the iodine memory effect, the mass spectrum of

$$C_5H_5Mo(CO)_3Cl$$

exhibits ions clearly arising from the three binuclear halides

$$[C_5H_5Mo(CO)Cl]_2, \quad [C_5H_5Mo(CO)I]_2,$$

and $[C_5H_5Mo(CO)]_2ICl$. The mass spectrum of the π-indenyl derivative $C_9H_7Mo(CO)_3I$ [38] (originally [53] formulated as $C_9H_7Mo(CO)_2I$) indicated that this compound was converted largely into the binuclear derivative [$C_9H_7MoI_2$]$_2$ formulated as *18*; the ions in the mass spectrum of this complex undergo competing losses of C_9H_7 and I fragments.

Several processes of interest occur in the mass spectrum of the π-allyl derivative $C_3H_5Fe(CO)_3I$ [52]. The series of ions $C_3H_5Fe(CO)_nI^+$ ($n = 3$, 2, 1, and 0) resulting from the stepwise loss of carbonyl groups is observed. However cleavage of the π-allyl and/or iodide ligands can compete effectively with loss of carbonyl groups giving rise to the additional series of ions $Fe(CO)_nI^+$ ($n = 2$, 1, and 0), $C_3H_5Fe(CO)_n^+$ ($n = 3, 2, 1$, and 0), and $Fe(CO)_n^+$ ($n = 2$, 1, and 0). A more unusual series of ions in the mass spectrum of $C_3H_5Fe(CO)_3I$ is $C_2H_2Fe(CO)_n^+$ ($n = 3$, 2, 1, and 0) formed apparently by elimination of methyl iodide; this process must involve a hydrogen shift.

17   *18*

The mass spectra of some metal nitrosyl halides have been studied [54]. The halides [$Fe(NO)_2X$]$_2$ (*19*: *10* = Cl or Br) which contain an iron-iron bond in addition to the metal halogen bridges exhibit stepwise loss of the nitrosyl groups to give the ions $Fe_2X_2^+$. By contrast, the halides [$Co(NO)_2X$]$_2$ (*20*: *10* = Cl, Br, or I) exhibit cleavage of the bimetallic system to $Co(NO)_2X^+$ which then undergoes elimination of nitrosyl groups and halogens. The mass spectrum of the cyclopentadienylmetal nitrosyl halide [$C_5H_5Mo(NO)I_2$]$_2$ (*21*) has also been investigated [52]. It does not exhibit a molecular ion. The highest m/e ion in its mass spectrum is $(C_5H_5)_2Mo_2(NO)_2I_2^+$ corresponding to elimination of I_2 from the molecular ion. In the proposed structure for [$C_5H_5Mo(NO)I_2$]$_2$ (*21*) the two non-bridging iodine atoms on different molybdenum atoms could be eliminated together as I_2 in a pyrolysis reaction. This would leave the compound $C_5H_5Mo(NO)I_2$ which could have structure *22* with a molybdenum-molybdenum bond and a favored 18-electron rare gas configuration for each molybdenum atom. The ion $(C_5H_5)_2Mo_2(NO)_2I_2^+$ undergoes competitive stepwise losses of nitrosyl groups and/or iodine atoms. The mass spectrum of [$C_5H_5Mo(NO)I_2$]$_2$ also exhibits the mononuclear ion $C_5H_5Mo(NO)I_2^+$ apparently formed by cleavage of the iodine bridges in *21*. This ion undergoes further fragmentation by loss of its nitrosyl group followed by elimination of C_5H_5, C_2H_2 and/or iodine.

108

In the case of the mass spectra of mononuclear terminal metal carbonyl hydrides such as $HMn(CO)_5$ [55] and $C_5H_5W(CO)_3H$ [19] hydrogen loss competes effectively with carbonyl loss even from the molecular ion. However, in the case of some polynuclear metal carbonyl hydrides with bridging hydrogen atoms, loss of hydrogen only occurs after most or all of the carbonyl groups are lost [56]. Thus in the mass spectrum of $HRe_3(CO)_{14}$ *(23)* hydrogen loss does not compete with the stepwise loss of carbonyl groups until the ion $HRe_3(CO)_2^+$ is reached; ions of the types $HRe_2(CO)_{9-n}^+$ are also observed as well as $Re_2(CO)_{10}^+$ apparently formed by a carbon monoxide transfer reaction. In the mass spectrum of

$$H_2Ru_4(CO)_{13}$$

the two hydrogen atoms are lost after loss of six carbonyl groups [57]. However, in the mass spectrum of $H_3Re_3(CO)_{12}$ losses of hydrogen

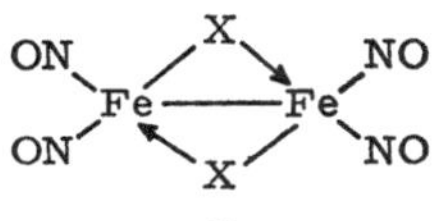

19 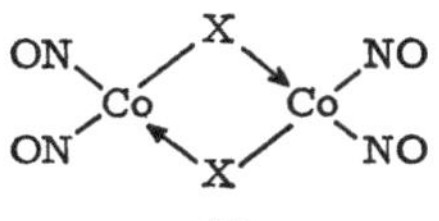   20 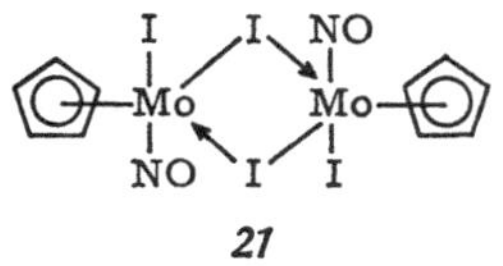   21

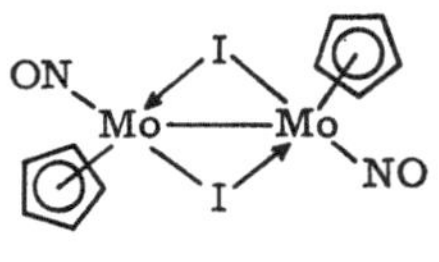

22

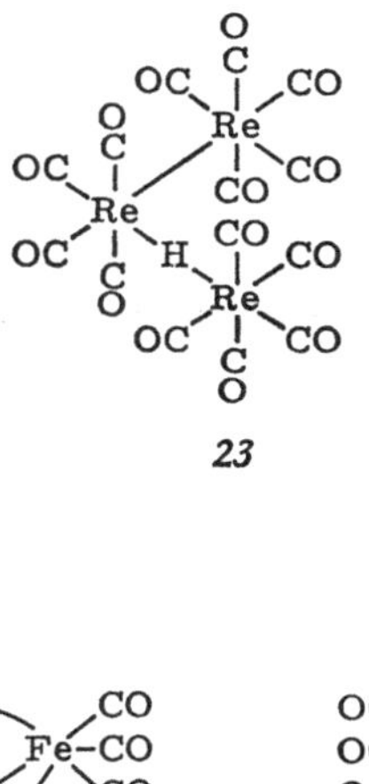

23

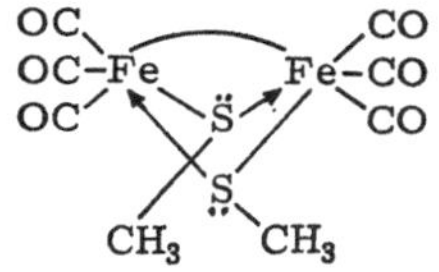

24 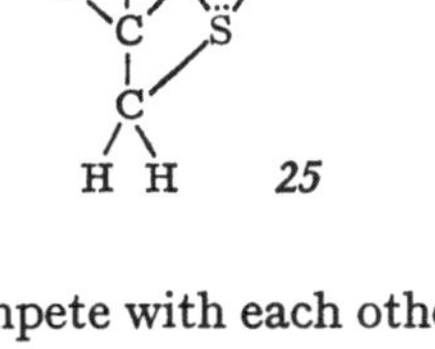   25 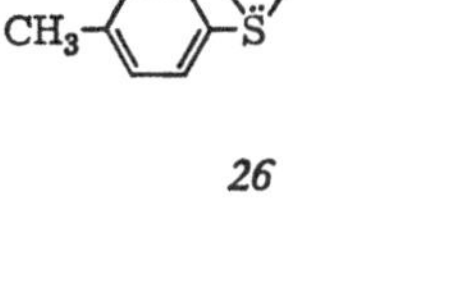   26

and carbon monoxide compete with each other [56]. In the mass spectrum of the manganese analogue, monometallic ions are predominantly observed apparently due to the expected greater weakness of manganese-manganese bonds than that of rhenium-rhenium bonds.

In the mass spectrum of $H_2Ru_4(CO)_{13}$ six carbonyl groups are lost followed by two hydrogens and then the remaining carbonyl groups [57]. Similar fragmentation patterns are noted for $H_4Ru_4(CO)_{12}$ [57] and the trinuclear osmium carbonyl hydrides $HOs_3(CO)_{10}Y$ [58]. (Y = H, OH, and OCH_3).

H. Mass Spectra of Alkylthio, Dialkylphosphido and Dialkylureylene Derivatives

The mass spectra of several compounds of the type $(RS)_2Fe_2(CO)_6$ (e.g. *24, 25,* and *26*) normally exhibits stepwise loss of carbonyl groups followed by cleavage of the carbon-sulfur bond [59]. In the case of $C_2H_4S_2Fe_2(CO)_6$ an ethylene fragment is eliminated from the $C_2H_4S_2Fe_2^+$ ion. In the mass spectra of $(RS)_2Fe_2(CO)_6$ compounds containing R groups with β-hydrogen atoms, (e.g. ethyl and *n*-butyl) neutral olefin fragments are eliminated from the carbonyl free ions $(RS)_2Fe_2^+$ resulting ultimately in the ion $Fe_2(SH)_2^+$ [51]. The manganese compound $H_2C_2S_2Mn_2(CO)_6$ *(27)* exhibits a similar fragmentation pattern; in this case losses of C_2H_2 and CO from the monocarbonyl ion $H_2C_2S_2Mn_2CO^+$ are competitive processes [60]. The mass spectra of certain bis(trifluoromethyl)ethylene-dithiolate derivatives such as $C_5H_5MS_2C_2(CF_3)_2$ (M = Co, Rh, and Ir), $[C_5H_5MoS_2C_2(CF_3)_2]_2$, and $C_5H_5W[S_2C_2(CF_3)_2]_2$ exhibit similar facile cleavage of carbon-sulfur bonds resulting in the elimination of neutral C_4F_6 fragments [61]. Cleavage of carbon-sulfur bonds predominates in the mass spectrum of the 2,5-dithiahexane complex $C_4H_{10}S_2W(CO)_4$ once all carbonyl groups are lost [59].

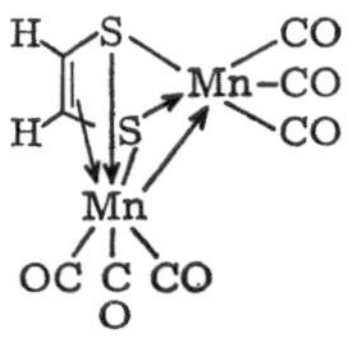

27

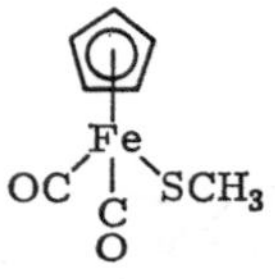

28

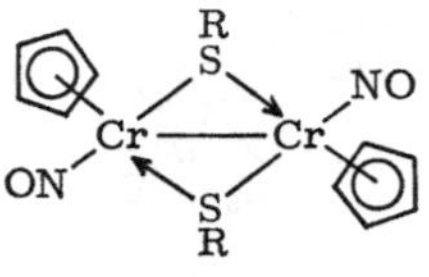

29

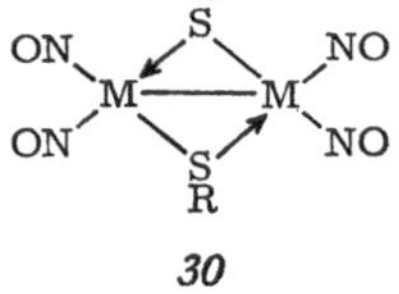

30

In the mass spectra of compounds with bridging alkylthio groups such as those discussed above, cleavage of the carbon-sulfur bond occurs before cleavage of the metal-sulfur bond. However, in the case of the mass spectrum of the compound $CH_3SFe(CO)_2C_5H_5$ *(28)* a similar cleavage of the carbon-sulfur bond in the carbonyl-free ion $C_5H_5FeSCH_3^+$ is not observed [59]. Instead the ion $C_5H_5FeSCH_3^+$ undergoes dehydrogenation to give the ion $C_6H_6SFe^+$ or hydrogen sulfide elimination to give the ion $C_6H_6Fe^+$. Again this behavior is consistent with the expected weaker bonding of a metal atom to a terminal sulfur atom with only one metal-sulfur bond than to a bridging sulfur atom with two (or sometimes more) metal-sulfur bonds.

The mass spectra of cyclopentadienylchromium nitrosyl derivatives of the type $[C_5H_5CrNOSR]_2$ *(29: R $= CH_3$ or C_6H_5)* [62] exhibit stepwise loss of the nitrosyl groups followed by carbon-sulfur bond cleavage with loss of the alkyl groups. The resulting $[C_5H_5CrS]_2^+$ ion then loses a neutral CrS_2 fragment to give the $(C_5H_5)_2Cr^+$ ion. The mass spectra of [RSM $(NO)_2]_2$ *(30: M $=$ Fe or Co; R $= C_2H_5$ or n-C_4H_9)* exhibit stepwise loss of nitrosyl groups followed by olefin elimination [54].

Reactions of tetraalkylbiphosphines with various metal carbonyls can give either metal carbonyl complexes of the tetraalkylbiphosphine without rupture of the phosphorus-phosphorus bond or metal carbonyl derivatives with bridging dialkylphosphido groups with rupture of the phosphorus-phosphorus bond [63]. The mass spectra of both types of compounds have been investigated [64]. The mass spectra of metal carbonyl complexes of tetramethylbiphosphine of the types $(CO)_nMP$-$(CH_3)_2P(CH_3)_2M(CO)_n$ (M $=$ Cr, Mo, and W, $n = 5$; M $=$ Fe, $n = 4$) exhibit the usual stepwise losses of carbonyl groups; in addition cleavage of the phosphorus-phosphorus bond competes with loss of carbonyl groups giving rise to ions of the type $(CH_3)_2PM(CO)_n^+$. The facile phosphorus-phosphorus bond cleavage in compounds of this type is consistent with the weakness of this bond as shown by the relatively long phosphorus-phosphorus bond length [65] in $(CO)_3NiP(C_6H_5)_2P(C_6H_5)_2Ni(CO)_3$ indicated by an X-ray crystallographic study. The mass spectra of the dimethylphosphido derivatives $[(CH_3)_2PM(CO)_n]_2$ (M $=$ Cr, Mo, and W, $n = 4$; M $=$ Fe, $n = 3$) exhibit only binuclear ions; the usual stepwise loss of carbonyl groups is observed. A comparison between the mass spectra of the chromium and manganese derivatives of the type

$$[(CH_3)_2PM(CO)_4]_2$$

shows little difference except for the presence of an ion of the type $(CH_3)_2PM^+$ in the manganese derivative but not the chromium derivative despite the fact that the two compounds have different structures: the chromium derivative but not the manganese derivative has a metal-metal bond. The mass spectra of the halides

$[(CH_3)_2PFe(CO)_3X]_2$ (*31*: $X = Cl$, Br, or I)

exhibit no parent ion or fragment ions containing both phosphorus and halogen. The iodide $[(CH_3)_2PFe(CO)_3I]_2$ exhibits a family of ions $Fe_2(CO)_nI_2^+$ ($0 \leq n \leq 6$) suggestive of pyrolysis in the mass spectrometer to give the iron carbonyl iodide $[Fe(CO)_3I]_2$ (*32*) which is otherwise unknown but would have a structure similar to that of the $[Fe(CO)_3SR]_2$ derivatives such as *34*.

Reactions of $Fe_3(CO)_{12}$ with alkyl isocyanates or alkyl azides give the N,N' dialkylureylene-diiron hexacarbonyls $(RN)_2COFe_2(CO)_6$ (*33*); mass spectroscopy was used for the initial identification of these compounds [66]. The mass spectrum of the diphenyl derivative

$(C_6H_5N)_2COFe_2(CO)_6$ (*33*: $R = C_6H_5$)

has been investigated in somewhat greater detail [67]. The molecular ion in this mass spectrum first loses stepwise its seven carbonyl groups giving the ion $(C_6H_5N)_2Fe_2^+$. The ion

$(C_6H_5N)_2COFe^+$

is more than twice as abundant as any other ions of the type $(C_6H_5N)_2Fe_2(CO)_n^+$ suggesting that the last carbonyl group is lost with much greater difficulty than the first six carbonyl groups. Furthermore the ion $(C_6H_5N)_2COFe_2^+$ besides losing its last carbonyl group can also fragment by the following three processes all supported by the presence of appropriate metastable ions:

1. Elimination of an HNCO fragment:

$$(C_6H_5N)_2COFe_2^+ \rightarrow C_6H_5NC_6H_4Fe_2^+ + HNCO$$

2. Elimination of an FeNCO fragment:

$$(C_6H_5N)_2COFe_2^+ \rightarrow (C_6H_5)_2FeN^+ + FeNCO$$

3. Elimination of a phenyl (iso)cyanide fragment:

$$(C_6H_5N)_2COFe_2^+ \rightarrow C_6H_5NFe_2O^+ + C_6H_5NC$$

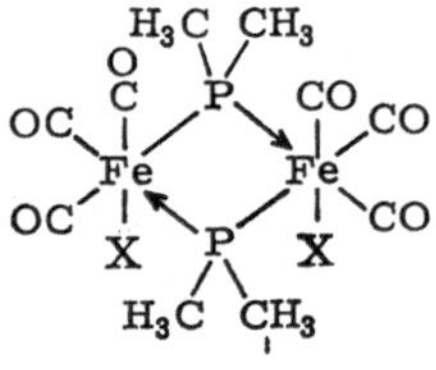

31 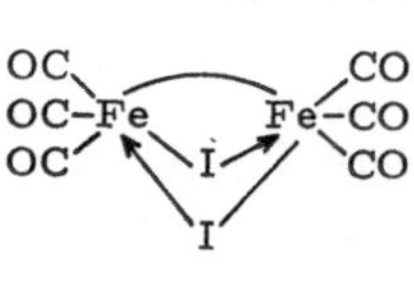   *32* 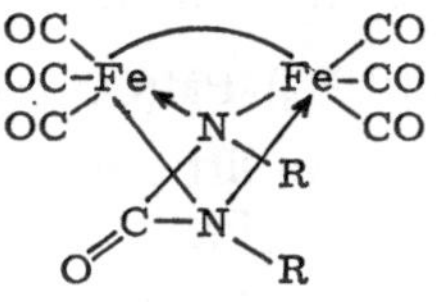   *33*

No evidence is available which indicates whether the neutral fragment eliminated in (1) is cyanic, isocyanic, or even fulminic acid or whether the neutral fragment eliminated in (3) is phenyl isocyanide or benzonitrile. These features of the mass spectrum of

$(C_6H_5N)_2COFe_2(CO)_6$

support the proposed structure *33* ($R = C_6H_5$) in which one of the seven carbonyl groups is not bonded to an iron atom but instead bridges two nitrogen atoms.

I. Mass Spectra of Fluorocarbon Derivatives of Transition Metals

The first mass spectral studies on fluorocarbon derivatives of metal carbonyls were reported in 1961 [68]; the octafluorotetramethylene derivative $C_4F_8Fe(CO)_4$ *(34)* and the octafluoro-1,3-cyclohexadiene

34   *35*   *36*

derivative $C_6F_8Fe(CO)_3$ *(35)* were included in this early work. In the mass spectra of both *34* and *35* stepwise losses of carbon monoxide occur followed by elimination of an FeF_2 fragment. Elimination of neutral metal fluoride fragments after loss of carbonyl groups was later shown by metastable ion evidence to be a rather general process in the mass spectra of fluorocarbon derivatives of metal carbonyls and cyclopentadienyls [69, 70]. Elimination of neutral FeF_2 fragments was observed in the mass spectra of the diverse iron complexes $C_2F_4S_2Fe_2(CO)_6$, $C_4F_6S_2Fe_2(CO)_6$, $C_6F_5Fe(CO)_2C_5H_5$, $3,4-H_2C_6F_3Fe(CO)_2C_5H_5$, $p-CF_3C_6F_4Fe(CO)_2C_5H_5$, $C_{14}H_{14}F_6Fe(CO)_3$, $C_3F_7Fe(CO)_4I$, and $C_2F_5Fe(CO)_2C_5H_5$. Elimination of a neutral CoF_2 fragment was observed [71] in the mass spectrum of $[(CF_3)_2C_2S_2CoCO]_3$. Elimination of a neutral C_5H_5CoF fragment was observed [69] to occur in the mass spectrum of the fluorinated bicyclo-[2,2,2] octatriene ("barrelene") derivative $C_5H_5CoC_{14}H_{14}F_6$ *(36)*. Elimination of neutral HF fragments frequently occur in the mass spectra of organic fluorine compounds [72]; such HF eliminations have been observed in the mass spectra of the fluorinated organometallic derivatives $C_6F_5Fe(CO)_2C_5H_5$, $3,4-H_2C_6F_3Fe(CO)_2C_5H_5$, $p-CF_3C_6F_4Fe(CO)_2C_5H_5$, $C_{14}H_{14}F_6Fe(CO)_3$, and $C_5H_5CoC_{14}H_{14}F_6$ [69].

Formation of neutral metal fluoride fragments as described above clearly requires a shift of fluorine from carbon to a metal atom. Similar fluorine shift processes can also account for some metal fluoride *ions* observed in the mass spectra of transition-metal-fluorocarbon derivatives. Thus many fluorocarbon derivatives of the cyclopentadienylmetal carbonyls exhibit the ions $C_5H_5MF^+$ which clearly must be formed by a fluorine shift process. In the mass spectra of the pentafluoropropenyl derivatives $CF_3CF=CFM(CO)_2C_5H_5$ (M = Fe or Ru) the ions $C_5H_5MF^+$ are clearly formed by elimination of a neutral C_3F_4 fragment from the carbonyl-free ions $C_3F_5MC_5H_5^+$ [70]. Elimination of a neutral CF_2 fragment from the ion $C_3F_5Re^+$ gives the ion $C_2F_3Re^+$ which could be $C_2F_2ReF^+$ [70]. The ion $CF_3MoC_5H_5^+$ in the mass spectrum of

$CF_3Mo(CO)_3C_5H_5$

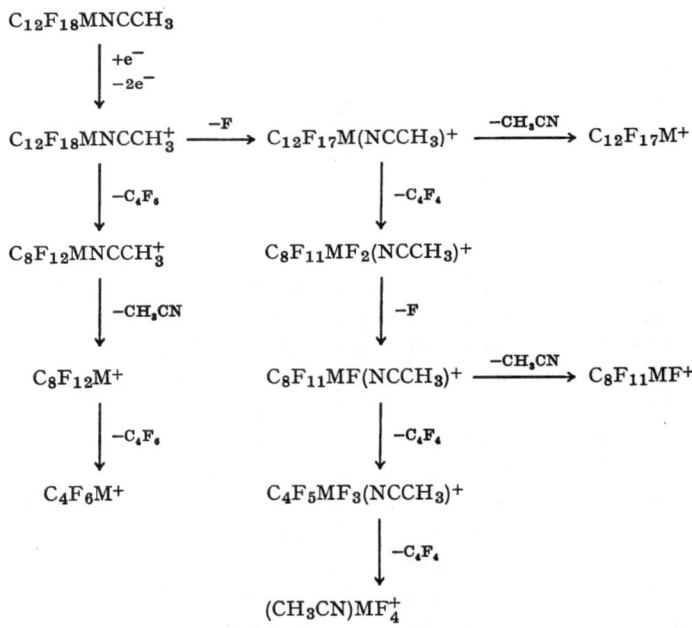

Fig. 3. General features of the fragmentation pattern involving the metal ions in the compounds $(CF_3C\equiv CCF_3)_3M(NCCH_3)$ (M = Mo and W)

appears to lose CF_2 with a fluorine shift to give the ion $C_5H_5MoF^+$ [73]. The mass spectra of $C_3F_7Mo(CO)_3C_5H_5$ and

$C_3F_7COW(CO)_3C_5H_5$

exhibit the ions $C_5H_5MF_2^+$ obviously formed by two fluorine shifts [73].

Hexafluorobutyne-2 reacts with the acetonitrile complexes $(CH_3 CN)_3M(CO)_3$ ($M = Mo$ and W) to form the very stable white crystalline derivatives $[(CF_3)_2C_2]_3MNCCH_3$ ($M = Mo$ and W). The fragmentation patterns (Fig. 3) exhibit an interesting sequence of C_4F_4 elimination combined with other processes [73]. The molecular ions in these mass spectra first appear to lose one fluorine atom giving the ions

$C_{12}F_{17}MNCCH_3^+$.

This ion then appears either to lose acetonitrile giving

$C_{12}F_{17}M^+$

or to lose a C_4F_4 fragment with shift of two fluorine atoms to the metal giving $C_8F_{11}MF_2(NCCH_3)^+$. The ion $C_8F_{11}MF_2(NCCH_3)^+$ then loses fluorine giving $C_8F_{11}MF(NCCH_3)^+$. This latter ion then appears to lose acetonitrile giving $C_8F_{11}MF^+$ or to lose C_4F_4 with shift of two more fluorine atoms to the metal atom giving $C_4F_5MF_3(NCCH_3)^+$. Another similar C_4F_4 elimination from the ion $C_4F_5MF_3(NCCH_3)^+$ followed by a fluorine shift gives the novel ion $CH_3CNMF_4^+$, a major metal-containing ion in the mass spectra of both the molybdenum and tungsten derivatives.

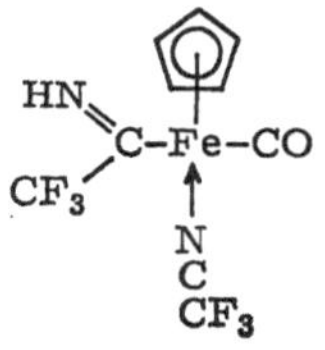

37

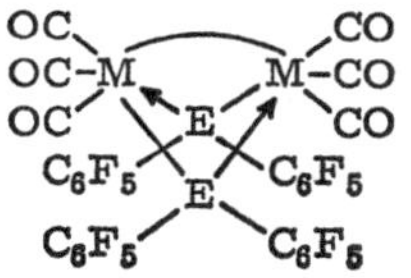

38

Trifluoroacetonitrile reacts with $CH_3Fe(CO)_2C_5H_5$ under pressure to give an unusual black trifluoroacetimino complex

$CF_3C(NH)Fe(CO)(NCCF_3)(C_5H_5)$ *(37)* [74].

The parent ion in the mass spectrum of this complex first loses its carbonyl group giving the ion $CF_3C(NH)Fe(NCCF_3)(C_5H_5)^+$ which can then undergo either loss of F or loss of CF_3CN; in the latter case the ion $CF_3C(NH)FeC_5H_5^+$ is formed. This ion then undergoes loss of a neutral CF_2 fragment to give $C_5H_5FeF(CNH)^+$ which then loses HCN to give $C_5H_5FeF^+$.

The mass spectra of several bis(pentafluorophenyl)phosphido and bis(pentafluorophenyl)arsenido derivatives of iron and ruthenium carbonyls of the type $[(C_6F_5)_2EM(CO)_3]_2$ (*38*: E = P or As; M = Fe or Ru) have been investigated [75,76]. Stepwise losses of carbonyl groups followed by elimination of C_6F_5FeF or MF_2 (M = Fe or Ru) fragments are observed.

J. Mass Spectra of π-Cyclopentadienyl Derivatives without Carbonyl Ligands

The first fragmentation process of the molecular ion in the mass spectra of π-cyclopentadienylmetal carbonyl derivatives is generally the loss of carbonyl groups [19]. Other processes of interest are generally only observed after all carbonyl groups are lost. The mass spectra of π-cyclopentadienyl derivatives without carbonyl groups are of interest because a greater variety of fragmentation processes occur with the molecular ion. In the mass spectra of most π-cyclopentadienyl derivatives without carbonyl ligands the molecular ion is the strongest metal-containing ion in the mass spectrum in contrast to the mass spectra of π-cyclopentadienylmetal carbonyl derivatives where the intensity of the molecular ion is always low compared with that of other metal-containing ions.

The relative intensities of the $C_5H_5M^+$ and MQ^+ ions in the mass spectra of the C_5H_5MQ compounds provide a useful qualitative indication of the relative strengths of the metal-Q bonds and the metal- π-cyclopentadienyl bonds [77]. The ratio $[MQ^+]/[C_5H_5M^+]$ for C_5H_5MQ derivatives containing iron or metals to the left of iron in the periodic table such as $C_5H_5MC_7H_7$ (M = V or Cr), $C_5H_5MnC_6H_6$, $C_5H_5FeC_9H_7$, and $C_5H_5FeC_4H_4N$ ranges from 0.0037 to 0.16 indicating a high abundance of $C_5H_5M^+$ relative to MQ^+. This suggests the high stability of the metal-

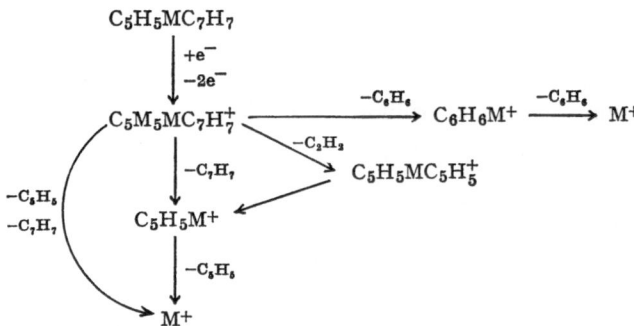

Fig. 4. Fragmentation scheme of $C_5H_5MC_7H_7$ compounds

π-cyclopentadienyl bond as compared with the metal-Q bond. This agrees with other observations on the stability of bonds between metal atoms and π-cyclopentadienyl rings as compared with other π-bonded ligands.

π-Cyclopentadienyl derivatives of nickel and palladium behave differently from those of iron and metals to the left of iron in the Periodic Table. The fragmentation patterns of the C_5H_5MQ compounds

$C_3H_5PdC_5H_5$, $C_{10}H_{12}OCH_3PdC_5H_5$, and $C_5H_5NiC_{13}H_{17}$

indicate that the π-C_5H_5 ring is lost at least as readily as the Q ligand from the parent ion. If the Q ligand contains a hydrogen atom attached to an sp^3 carbon atom (e.g. $C_{10}H_{12}OCH_3$ and $C_{13}H_{17}$ [tetramethylcyclo-pentadienylbutenyl]), the π-bonded C_5H_5 ligand abstracts this hydrogen atom and is eliminated as a C_5H_6 (cyclopentadiene) fragment. In the case of $C_3H_5PdC_5H_5$ where no such hydrogen atom is available, the π-bonded C_5H_5 ring appears to be eliminated as a neutral C_5H_5 fragment.

The mass spectra of the mixed π-cyclopentadienyl-π-cyclohepta-trienyl derivatives $C_5H_5MC_7H_7$ (39: M = V or Cr) exhibit sufficient metastable ions to provide reasonable evidence in support of four fragmentation pathways (Fig. 4) from the molecular ions to the "bare" metal ions [77,78]. In the first fragmentation pathway the molecular ion first loses its C_7H_7 ring giving the ion $C_5H_5M^+$ which then loses its C_5H_5 ring giving the bare metal ion M^+. In the second fragmentation pathway the molecular ion loses both its C_5H_5 and C_7H_7 rings in one step to give the bare metal ion M^+. In the third fragmentation pathway the molecular ion first loses a neutral C_2H_2 fragment giving the ion $C_5H_5MC_5H_5^+$; this ion then loses successively its two C_5H_5 groups giving the bare metal ion M^+.

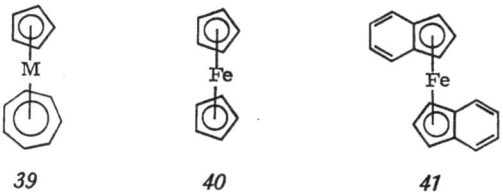

39   40   41

The fourth fragmentation pathway from the molecular ions

$C_5H_5MC_7H_7^+$

down to the bare metal ions M^+ is particularly unusual. In this pathway the molecular ions go to the bare metal ions by successive losses of two C_6H_6 fragments. Metastable ions for both losses of the C_6H_6 fragments for both the vanadium and chromium complexes are observed thereby strongly supporting the existence of this pathway. The ability for $C_5H_5MC_7H_7^+$ ions without any C_6H_6 rings to fragment by loss of neutral C_6H_6 fragments is an interesting demonstration of the stability of neutral C_6H_6 fragments probably corresponding to benzene.

The mass spectrum of the π-pyrrolyl derivative $C_5H_5FeC_4H_4N$ *(40)* exhibits a very strong molecular ion peak [77]. The following four fragmentation processes of the molecular ion were confirmed by observation of appropriate metastable ions:

1. Elimination of C_2H_2

$$C_5H_5FeC_4H_4N^+ \rightarrow C_5H_5FeC_2H_2N^+ + C_2H_2$$

2. Elimination of HCN

$$C_5H_5FeC_4H_4N^+ \rightarrow C_5H_5FeC_3H_3^+ + HCN$$

3. Elimination of C_2H_2N

$$C_5H_5FeC_4H_4N^+ \rightarrow C_5H_5FeC_2H_2^+ + C_2H_2N$$

4. Elimination of the entire C_4H_4N ligand

$$C_5H_5FeC_4H_4N^+ \rightarrow C_5H_5Fe^+ + C_4H_4N$$

In addition the ion $C_9H_9N^+$ is observed which may be formed by elimination of a neutral iron atom from the molecular ion.

The mass spectra of several π-indenyl derivatives have been investigated [38]. Fig. 5 shows the predominant features of the fragmentation scheme of $(C_9H_7)_2Fe$ *(41)*. The parent ion in this mass spectrum first loses one of its indenyl ligands giving the ion $C_9H_7Fe^+$. This ion then

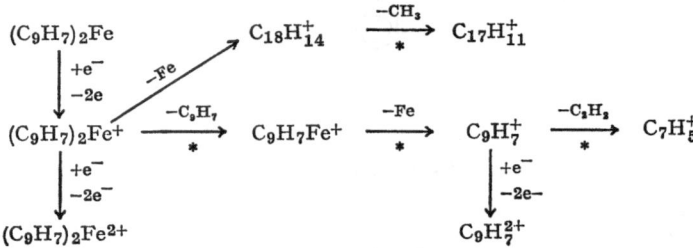

* Metastable ions were observed for these processes.

Fig. 5. Fragmentation scheme of $(C_9H_7)_2Fe$

loses its iron atom giving the indenyl ion $C_9H_7^+$. The only other iron ion present in significant abundance is the doubly charged molecular ion $(C_9H_7)_2Fe^{2+}$. The bare iron ion Fe^+ is only observed in low abundance.

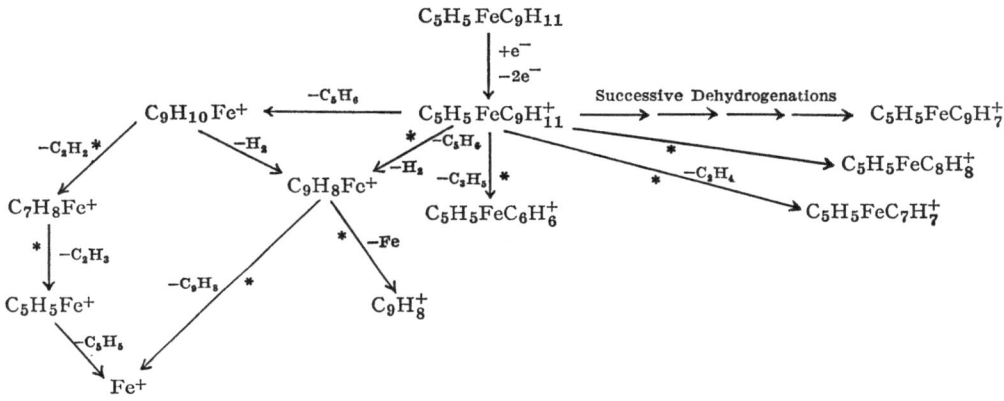

Fig. 6. Fragmentation scheme of $C_5H_5FeC_9H_{11}$

Apparently the indenyliron ion $C_9H_7Fe^+$ fragments almost exclusively by elimination of a neutral iron atom according to the following equation:

$$C_9H_7Fe^+ \rightarrow C_9H_7^+ + Fe$$

This contrasts with the behavior of the corresponding cyclopentadienyliron ion $C_5H_5Fe^+$ which fragments mainly by elimination of a neutral C_5H_5 fragment according to the following equation:

$$C_5H_5Fe^+ \rightarrow Fe^+ + C_5H_5$$

The ion $C_{18}H_{14}^+$ appears to arise by elimination of an iron atom from the molecular ion with concurrent coupling of the two indenyl ligands.

The molecular ion of π-cyclopentadienyl-π-indenyliron (benzoferrocene, 42) $C_5H_5FeC_9H_7$ can lose either its indenyl (C_9H_7) ligand giving the ion $C_5H_5Fe^+$ or its cyclopentadienyl (C_5H_5) ligand giving the ion $C_9H_7Fe^+$. The ratio of the relative abundance of $C_9H_7Fe^+$ to that of $C_5H_5Fe^+$ is only 0.071 indicating that the tendency for the parent ion to lose its π-indenyl ligand is much greater than the tendency for the parent ion to lose its π-cyclopentadienyl ligand. Further fragmentation of $C_5H_5Fe^+$ results in the elimination of a C_5H_5 fragment giving Fe^+ whereas further fragmentation of $C_9H_7Fe^+$ results in the elimination of a neutral iron atom giving $C_9H_7^+$.

Hydrogenation of $C_5H_5FeC_9H_7$ (42) at atmospheric pressure in the presence of a palladium catalyst has been shown to give $C_5H_5FeC_9H_{11}$ (43) [79]. Fig. 6 depicts the principal features of the fragmentation scheme in the mass spectrum of $C_5H_5FeC_9H_{11}$. The molecular ion undergoes the following four processes indicated by metastable ion analysis:

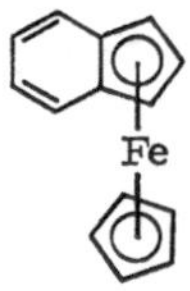

42 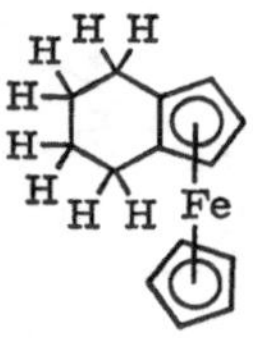   *43*

1. Elimination of CH_3

$$C_5H_5FeC_9H_{11}^+ \rightarrow C_5H_5FeC_8H_8^+ + CH_3$$

This process must necessarily involve a hydrogen shift.

2. Elimination of C_2H_4

$$C_5H_5FeC_9H_{11}^+ \rightarrow C_5H_5FeC_7H_7^+ + C_2H_4$$

Two adjacent sp^3 carbon atoms in the six-membered ring of the C_9H_{11} ligand can be eliminated as ethylene.

3. Elimination of C_3H_5

$$C_5H_5FeC_9H_{11}^+ \rightarrow C_5H_5FeC_6H_6^+ + C_3H_5$$

The $C_5H_5FeC_6H_6^+$ ion formed in this reaction has also been isolated in the form of stable salts.

4. Elimination of $C_5H_6 + H_2$

$$C_5H_5FeC_9H_{11}^+ \rightarrow C_9H_8Fe^+ + C_5H_6 + H_2$$

The presence of an appreciable abundance of $C_9H_{10}Fe^+$ also suggests elimination of the π-cyclopentadienyl ligand as C_5H_6 from the parent ion.

120

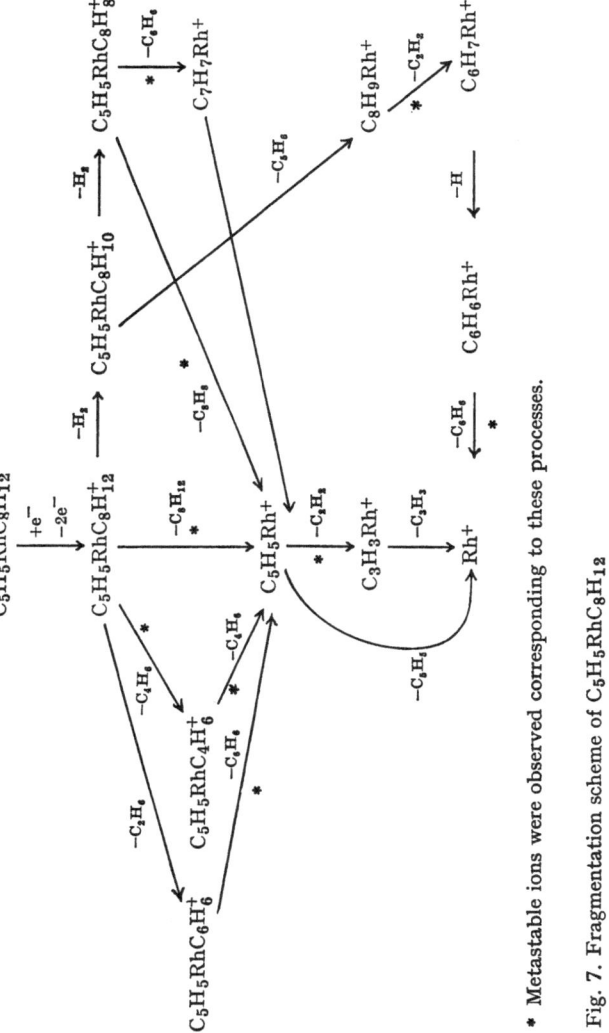

Fig. 7. Fragmentation scheme of $C_5H_5RhC_8H_{12}$

By use of the cyclononatetraenide anion the compound $C_5H_5FeC_9H_9$, an intermediate hydrogenation product of $C_5H_5FeC_9H_7$, is obtained [38]. The molecular ion in the mass spectrum of $C_5H_5FeC_9H_9$ undergoes dehydrogenation as far as $C_5H_5FeC_9H_7^+$ and C_2H_2 elimination as far as $(C_5H_5)_2Fe^+$. Other processes observed in the mass spectra of $C_5H_5FeC_9H_9$ are similar to those discussed for other organometallic compounds.

The mass spectra of some cyclopentadienylrhodium olefin complexes have been investigated [80]. The molecular ion in the mass spectrum of the ethylene complex $C_5H_5Rh(C_2H_4)_2$ undergoes fragmentation to the bare metal ion by the following sequence:

1. $\qquad C_5H_5Rh(C_2H_4)_2^+ \rightarrow C_5H_5RhC_2H_4^+ + C_2H_4$

2. $\qquad C_5H_5RhC_2H_4^+ \rightarrow C_5H_5Rh^+ + C_2H_4$

3. $\qquad C_5H_5Rh^+ \rightarrow Rh^+ + C_5H_5$

The ethylene groups are thus lost stepwise much like the carbonyl groups in metal carbonyls. The relative abundance of the molecular ion in the mass spectrum of $C_5H_5Rh(C_2H_4)_2$ is only about 10% of that of the strongest rhodium ion $C_5H_5Rh^+$ similar to the relative abundance of the molecular ions in similar metal carbonyl derivatives.

As might be expected the mass spectrum of the 1,5-cyclooctadiene complex $C_5H_5RhC_8H_{12}$ *(44)* exhibited a much more complex fragmentation pattern (Fig. 7). Several pathways appear to be possible in going from the parent ion to the ion $C_5H_5Rh^+$, which can then fragment to a bare rhodium ion Rh^+ by cleavage of its C_5H_5 ring. Thus, the parent ion

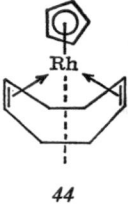

44

$C_5H_5RhC_8H_{12}$ can form the ion $C_5H_5Rh^+$ by loss of the C_8H_{12} ligand in a single step. Alternatively the parent ion can lose the elements of ethane (C_2H_6 or $2\,CH_3$) to form the ion $C_5H_5RhC_6H_6^+$, which can then lose its C_6H_6 ligand in a further step forming $C_5H_5Rh^+$. The molecular ion $C_5H_5RhC_8H_{12}^+$ may also undergo dehydrogenation forming

$C_5H_5RhC_8H_8^+$

which can then lose a C_8H_8 fragment forming $C_5H_5Rh^+$. Alternatively the ion $C_5H_5RhC_8H_8^+$ can lose a neutral C_6H_6 fragment forming an ion $C_7H_7Rh^+$, which then appears to form $C_5H_5Rh^+$ with expulsion of a neutral C_2H_2 fragment. Still another possible route from $C_5H_5RhC_8H_{12}^+$ to $C_5H_5Rh^+$ involves the intermediate ion $C_5H_5RhC_4H_6^+$;

this suggests that under certain conditions 1,5-cyclooctadiene can break down into butadiene units in the mass spectrometer as indicated by the dotted lines in *44*.

The mass spectrum of $C_5H_5RhC_8H_{12}$ also exhibits the ion $C_8H_9Rh^+$ possibly formed by elimination of a neutral cyclopentadiene (C_5H_6) fragment from the ion $C_5H_5RhC_8H_{10}^+$. The ion $C_8H_9Rh^+$ can fragment to the bare rhodium ion Rh^+ by successive losses of C_2H_2, H, and C_6H_6. The fact that this further degradation of $C_8H_9Rh^+$ ion bypasses the $C_5H_5Rh^+$ step suggests that $C_8H_9Rh^+$ is not $C_5H_5RhC_3H_4^+$ still containing a π-cyclopentadienyl ring.

These observations suggest that in different fragmentation modes of the molecular ion of $C_5H_5RhC_8H_{12}$ *(44)* either the C_5H_5 ring or the C_8H_{12} ring may be lost. This indicates that the fragmentation pattern of $C_5H_5RhC_8H_{12}$ in the mass spectrum is intermediate between those of typical π-cyclopentadienyl derivatives of iron, manganese, chromium, vanadium, and their heavier congeners and those of π-cyclopentadienyl derivatives of nickel and palladium. In the former cases the ligands other than π-cyclopentadienyl are lost most readily in the mass spectrum resulting in relatively high abundances of the $C_5H_5M^+$ ions. In the latter cases an important fragmentation pathway of the molecular ion is loss of the π-cyclopentadienyl ring as C_5H_6 resulting in relatively high abundances of cyclopentadienyl-free metal ions.

The mass spectra of several π-cyclopentadienyl derivatives of nickel, palladium, and platinum have also been investigated [77]. The molecular ion in the mass spectrum of the π-cyclopentadienylplatinum derivative $(CH_3)_3PtC_5H_5$ can fragment either by loss of the cyclopentadienyl ring forming $(CH_3)_3Pt^+$ or by loss of the methyl groups forming $CH_3PtC_5H_5^+$ and $C_5H_5Pt^+$. The molecular ion in the mass spectrum of the π-allyl-palladium derivative $C_3H_5PdC_5H_5$ appears to fragment either by loss

45   46

of the π-allyl group as allene giving $C_5H_6Pd^+$ or by loss of the π-cyclopentadienyl group as C_5H_5 giving $C_3H_5Pd^+$. The ion $C_5H_5Pd^+$ was not detected in appreciable quantities; the mass spectrum of $C_3H_5PdC_5H_5$ thus differs from that of most π-cyclopentadienyl derivatives. However chemical studies [81] have shown the π-cyclopentadienyl ring to be removed from $C_3H_5PdC_5H_5$ more readily than the π-allyl group. The molecular ion in the mass spectrum of the π-cyclopentadienylpalladium derivative $(C_{10}H_{12}OCH_3)PdC_5H_5$ *(45)* loses a C_5H_6 fragment forming the ion $C_{10}H_{11}OCH_3Pd^+$ which is much more abundant than the combined $C_5H_6Pd^+$ and $C_5H_5Pd^+$ ions; this suggests that in *45* the $C_{10}H_{12}OCH_3$ ligand is more tightly bonded to the palladium atom than the π-cyclopentadienyl ligand. Similarly in the mass spectrum of the π-cyclopentadienylnickel derivative $C_5H_5NiC_{13}H_{17}$ *(46)* the ion $NiC_{13}H_{16}^+$ formed by loss of the π-cyclopentadienyl ring as C_5H_6 from the molecular ion is more than twice as abundant as the ion $C_5H_5Ni^+$ formed by loss of the $C_{13}H_{17}$ ligand from the molecular ion. These data indicate than in *46* the substituted π-cyclobutenyl ligand is more strongly bonded to the nickel atom than the π-cyclopentadienyl ring.

K. References

[1] *King, R. B.:* Transition Metal Organometallic Chemistry: An Introduction. New York: Academic Press 1969.

[2] *Bird, C. W.:* Transition Metal Intermediates in Organic Synthesis. London: Logos Press 1967.

[3] *Friedman, L., A. P. Irsa,* and *G. Wilkinson:* J. Am. Chem. Soc. *77*, 3689 (1955).

[4] *Bruce, M. I.:* Advan. Organometal. Chem. *6*, 273 (1968).

[5] *Cais, M.,* and *M. S. Lupin:* Advan. Organometal. Chem., in press.

[6] *Budzikiewicz, H., C. Djerassi,* and *D. H. Williams:* Interpretation of Mass Spectra of Organic Compounds. San Francisco: Holden-Day, Inc. 1964.

[7] *Beynon, J. H.:* Mass Spectrometry and its Applications to Organic Chemistry. Amsterdam: Elsevier Publishing Company 1960.

[8] *Kiser, R. W.:* Introduction to Mass Spectrometry and its Applications. New York: Prentice Hall 1965.

[9] *Biemann, K.:* Mass Spectrometry: Organic Chemical Applications. New York: McGraw-Hill 1962.

[10] *Reed, R. I.:* Applications of Mass Spectrometry to Organic Chemistry. New York: Academic Press 1966.

[11] *Melton, C. E.:* In: Mass Spectrometry of Organic Ions; *F. W. McLafferty,* Ed. New York: Academic Press 1963.

[12] *Winters, R. E.,* and *R. W. Kiser:* Inorg. Chem. *3*, 699 (1964).

[13] — — Inorg. Chem. *4*, 157 (1965).

[14] — — J. Phys. Chem. *69*, 1618 (1965).

[15] *Dougherty, R. C.,* and *C. R. Weisenberger:* J. Am. Chem. Soc. *90*, 6570 (1968).

[16] *Beckey, H. D., H. Knöppel, G. Metzinger,* and *P. Schulze:* Advan. Mass Spectrometry *3*, 35—68 (1966).

[17] *Becconsall, J. K., B. E. Job,* and *S. O'Brien:* J. Chem. Soc. A 423 (1967).

[18] *Field, F. H.:* Accounts Chem. Res. *1*, 42 (1968).

[19] *King, R. B.:* J. Am. Chem. Soc. *90*, 1417 (1968).

[20] *Carrick, A.,* and *F. Glockling:* J. Chem. Soc. A 40 (1967).

[21] *Taubert, R.,* and *F. P. Lossing:* J. Am. Chem. Soc. *84*, 1523 (1962).

22) *Melton, C. E.*, and *W. H. Hamill:* J. Chem. Phys. *41*, 546 (1964).
23) *Schumacher, E.*, and *R. Taubenest:* Helv. Chim. Acta *47*, 1525 (1964).
24) *Müller, J.*, and *L. D'Or:* J. Organometal. Chem. *10*, 313 (1967).
25) *Johnson, B. F. G., J. Lewis, I. G. Williams*, and *J. M. Wilson:* J. Chem. Soc. A 341 (1967).
26) *King, R. B.:* J. Am. Chem. Soc. *88*, 2075 (1966).
27) *Svec, H. J.*, and *G. A. Junk:* J. Am. Chem. Soc. *89*, 2836 (1967).
28) *Robinson, B. H.*, and *W. S. Tham:* J. Chem. Soc. A 1784 (1968).
29) *Mays, M. J.*, and *R. N. F. Simpson:* J. Chem. Soc. A 1444 (1968).
30) *Johnson, B. F. G., R. D. Johnston*, and *J. Lewis:* J. Chem. Soc. A 2865 (1968).
31) *Schumacher, E.*, and *R. Taubenest:* Helv. Chim. Acta *49*, 1447 (1966).
32) *King, R. B.*, and *M. Ishaq:* to be published.
33) — Org. Mass Spectrometry *2*, 657 (1969).
34) — Chem. Commun. 986 (1967).
35) — Inorg. Chem. *5*, 2227 (1966).
36) — Chem. Commun. 436 (1969).
37) *Müller, J.*, and *M. Herberhold:* J. Organometal Chem. *13*, 399 (1968).
38) *King, R. B.:* Can. J. Chem. *47*, 559 (1969).
39) *Pignataro, S.*, and *F. P. Lossing:* J. Organometal. Chem. *10*, 531 (1967).
40) *King, R. B.:* paper presented at the First International Symposium on New Aspects of the Chemistry of Metal Carbonyls and Derivatives, Venice, Italy, September, 1968, Proceedings, paper E 7; Appl. Spectry. *23*, 536 (1969).
41) *Winters, R. E.*, and *R. W. Kiser:* J. Phys. Chem. *69*, 3198 (1965).
42) *Nakamura, A., P. J. Kim*, and *N. Hagihara:* J. Organometal. Chem. *3*, 7 (1965).
43) *King, R. B.*, and *F. G. A. Stone:* J. Am. Chem. Soc. *82*, 4557 (1960).
44) *Churchill, M. R.*, and *J. Wormald:* Chem. Commun. 1597 (1968).
45) *King, R. B.:* Inorg. Chem. *2*, 936 (1963).
46) — J. Am. Chem. Soc. *90*, 1412 (1968).
47) *Braterman, P. S.:* J. Organometal. Chem. *11*, 198 (1968).
48) *Lewis, J., A. R. Manning, J. R. Miller*, and *J. M. Wilson:* J. Chem. Soc. A 1663 (1966).
49) *Kiser, R. W., M. A. Krassoi*, and *R. J. Clark:* J. Am. Chem. Soc. *89*, 3653 (1967).
50) *Saalfeld, F. E., M. V. McDowell, S. K. Gondal*, and *A. G. MacDiarmid:* J. Am. Chem. Soc. *90*, 3684 (1968).
51) *Edgar, K., B. F. G. Johnson, J. Lewis, I. G. Williams*, and *J. M. Wilson:* J. Chem. Soc. A 379 (1967).
52) *King, R. B.:* Org. Mass Spectrometry *2*, 401 (1969).
53) —, and *M. B. Bisnette:* Inorg. Chem. *4*, 475 (1965).
54) *Johnson, B. F. G., J. Lewis, I. G. Williams*, and *J. M. Wilson:* J. Chem. Soc. A 338 (1967).
55) *Edgell, W. F.*, and *W. M. Risen:* J. Am. Chem. Soc. *88*, 5451 (1966).
56) *Smith, J. M., K. Mehner*, and *H. D. Kaesz:* J. Am. Chem. Soc. *89*, 1759 (1967).
57) *Johnson, B. F. G., R. D. Johnston, J. Lewis, B. H. Robinson*, and *G. Wilkinson:* J. Chem. Soc. A 2856 (1968).
58) —, *J. Lewis*, and *P. A. Kilty:* J. Chem. Soc. A 2859 (1968).
59) *King, R. B.:* J. Am. Chem. Soc. *90*, 1429 (1968).
60) —, and *C. A. Eggers:* Inorg. Chem. *7*, 1214 (1968).
61) —, and *M. B. Bisnette:* Inorg. Chem. *6*, 469 (1967).
62) *Preston, F. J.*, and *R. I. Reed:* Chem. Commun. 51 (1966).
63) *Hayter, R. G.:* Preparative Inorg. Reactions (*W. L. Jolly*, Ed.) *2*, 211—236 (1965).
64) *Johnson, B. F. G., J. Lewis, J. M. Wilson*, and *D. T. Thompson:* J. Chem. Soc. A 1445 (1967).

65) *Mais, R. H. B., P. G. Owston, D. T. Thompson,* and *A. M. Wood:* J. Chem. Soc. A 1744 (1967).

66) *Flannigan, W. T., G. R. Knox,* and *P. L. Pauson:* Chem. (London) 1094 (1967).

67) *King, R. B.:* Org. Mass Spectrometry 2, 381 (1969).

68) *Hoehn, H. H., L. Pratt, K. F. Watterson,* and *G. Wilkinson:* J. Chem. Soc. 2738 (1961).

69) *King, R. B.:* J. Am. Chem. Soc. *89,* 6368 (1967).

70) *Bruce, M. I.:* Org. Mass Spectrometry 2, 63 (1969).

71) *King, R. B.,* and *T. F. Korenowski:* Chem. Commun. 771 (1966).

72) *Budzikiewicz, H., C. Djerassi,* and *D. H. Williams:* Interpretation of Mass Spectra of Organic Compounds, p. 76. San Francisco: Holden-Day, Inc. 1964.

73) *King, R. B.:* Appl. Spectry. *23,* 137 (1969).

74) —, and *K. H. Pannell:* J. Am. Chem. Soc. *90,* 3984 (1968).

75) *Miller, J. M.:* J. Chem. Soc. A 828 (1967).

76) *Cooke, M., M. Green,* and *D. Kirkpatrick:* J. Chem. Soc. A 1507 (1968).

77) *King, R. B.:* Appl. Spectry. *23,* 148 (1969).

78) *Müller, J.,* and *P. Göser:* J. Organometal. Chem. *12,* 163 (1968).

79) *King, R. B.,* and *M. B. Bisnette:* Inorg. Chem. *3,* 796 (1964).

80) — J. Organometal Chem. *14,* P 19 (1968).

81) *Gubin, S. P., A. Z. Rubezhov, B. L. Winch,* and *A. N. Nesmeyanov:* Tetrahedron Letters 2881 (1964).

Received May 19, 1969

SPRINGER-VERLAG
BERLIN·HEIDELBERG·NEW YORK

Fortschritte der chemischen Forschung

Herausgeber: A. Davison, Cambridge, MA; M. J. S. Dewar, Austin, TX; K. Hafner, Darmstadt; E. Heilbronner, Basel; U. Hofmann, Heidelberg; K. Niedenzu, Lexington KY; Kl. Schäfer, Heidelberg; G. Wittig, Heidelberg

Schriftleitung: F. Boschke, Heidelberg

Band 12 / Heft 1

Mit 69 Abbildungen
184 Seiten. 1969
Geheftet DM 59,–
US $ 16.30

Organische Chemie

Inhalt: P. Zuman, Polarography in Organic Chemistry. – W. Steglich, Fortschritte in der Chemie der Oxazolinone-(5). — F. L. Breusch, Homologe und isomere Reihen.

Band 12 / Heft 2

Mit 75 Abbildungen
202 Seiten. 1969
Geheftet DM 54,–
US $ 14.90

Angewandte Chemie

Inhalt: G. Ohloff, Chemie der Geruchs- und Geschmacksstoffe. — H. Kölbel und P. Kurzendörfer, Konstitution und Eigenschaften von Tensiden. — H. Sackmann und D. Demus, Eigenschaften und Strukturen thermotroper kristallinflüssiger Zustände.

Band 12 / Heft 3

Mit 33 Abbildungen
151 Seiten. 1969
Geheftet DM 49,–
US $ 13.50

Organische Chemie

Inhalt: H. F. Ebel, Struktur und Reaktivität von Carbanionen und carbanionoiden Verbindungen. — M. Spiteller-Friedmann und G. Spiteller, Massenspektren von Steroiden.

Band 12 / Heft 4

Mit 37 Abbildungen
236 Seiten. 1969
Geheftet DM 59,–
US $ 16.30

Radiochemie

Inhalt: F. Weigel, Die Chemie des Promethiums. — K. H. Neeb und W. Gebauhr, Probleme der Kernbrennstoffanalyse. — F. Baumgärtner und H. Philipp, Die Wiederaufarbeitung von Uran-Plutonium-Kernbrennstoffen.

ISBN 978-3-540-04816-9   ISBN 978-3-540-36195-4 (eBook)
DOI 10.1007/978-3-540-36195-4

Titel-Nr. 7718